STAR 21

STRATEGIC TECHNOLOGIES FOR THE ARMY OF THE TWENTY-FIRST CENTURY

Board on Army Science and Technology
Commission on Engineering and Technical Systems
National Research Council

NATIONAL ACADEMY PRESS
Washington, D.C. 1992

NATIONAL ACADEMY PRESS • 2101 Constitution Avenue, N.W. • Washington, D.C. 20418

NOTICE: The project that is the subject of this report was approved by the Governing Board of the National Research Council, whose members are drawn from the councils of the National Academy of Sciences, the National Academy of Engineering, and the Institute of Medicine. The members of the committee responsible for the report were chosen for their special competences and with regard for appropriate balance.

This report has been reviewed by a group other than the authors according to procedures approved by a Report Review Committee consisting of members of the National Academy of Sciences, the National Academy of Engineering, and the Institute of Medicine.

This work is related to Department of the Army Contract DAAG29-85-C-0008 (CLIN 10). However, the contents do not necessarily reflect the positions or the policies of the Department of the Army or the U.S. government, and no official endorsement should be inferred.

National Research Council (U.S.). Board on Army Science and
 Technology.
   STAR 21 : strategic technologies for the army of the twenty-first
 century / Board on Army Science and Technology, Commission on
 Engineering and Technical Systems, National Research Council.
     p.   cm.
   Includes index.
   ISBN 0-309-04629-7 : $34.95
   1. Military engineering—United States. 2. United States. Army—
 Equipment. I. Title. II. Title: STAR twenty-one.
 UG23.N38   1992
 623'.0973—dc20   92-8945
                                                               CIP

Printed in the United States of America

Preface

The Assistant Secretary of the Army for Research, Development and Acquisition [ASA(RDA)] wrote to the Chairman of the Board on Army Science and Technology in March 1988 to request a study under the auspices of the National Research Council. The study's goal would be to assist the Army in improving its ability to incorporate advanced technologies into its weapons, equipment, and doctrine. The time period to be addressed by the study was specified to extend at least 30 years into the future. The three study objectives stated in the request were to (1) identify the advanced technologies most likely to be important to ground warfare in the next century, (2) suggest strategies for developing the full potential of these technologies, and (3) project implications for force structure and strategy of the technology changes.

The ASA(RDA) expressed the belief that the expert, independent advice provided by such a study would help the Army in selecting those strategic technologies that offer the greatest opportunity for increasing the effectiveness of forces in the field. The study would also assist the Army in designing current research and development strategies to ensure that such advanced technologies do become available for future Army applications.

To conduct the study, the National Research Council organized a Committee on Strategic Technologies for the Army (STAR) with nine science and technology groups and eight systems panels (Figure P-1). These were subordinated to a Science and Technology Subcommittee and an Integration Subcommittee, respectively. In addition, a Tech-

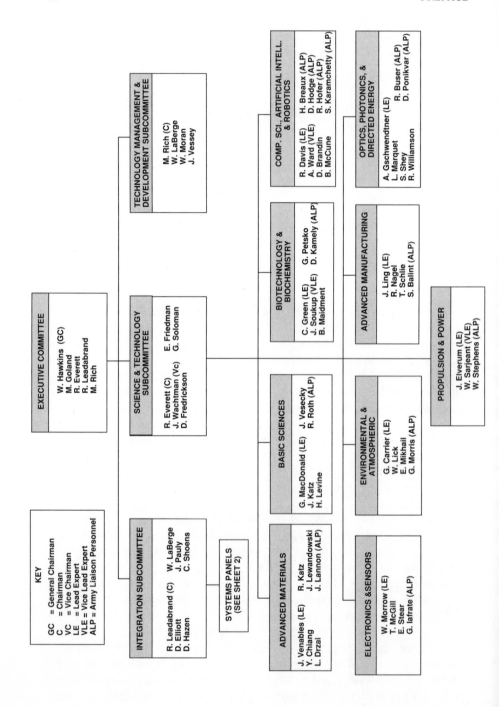

KEY

GC = General Chairman
C = Chairman
VC = Vice Chairman
LE = Lead Expert
VLE = Vice Lead Expert
ALP = Army Liaison Personnel

EXECUTIVE COMMITTEE

W. Hawkins (GC)
M. Goland
R. Everett
R. Leadabrand
M. Rich

TECHNOLOGY MANAGEMENT & DEVELOPMENT SUBCOMMITTEE

M. Rich (C)
W. LaBerge
W. Moran
J. Vessey

INTEGRATION SUBCOMMITTEE

R. Leadabrand (C)   W. LaBerge
D. Elliott   J. Pauly
D. Hazen   C. Shoens

SYSTEMS PANELS (SEE SHEET 2)

SCIENCE & TECHNOLOGY SUBCOMMITTEE

R. Everett (C)   E. Friedman
J. Wachtman (Vc)   G. Soloman
D. Fredrickson

COMP. SCI., ARTIFICIAL INTELL. & ROBOTICS

R. Davis (LE)   H. Breaux (ALP)
A. Ward (VLE)   D. Hodge (ALP)
D. Brandin   R. Hofer (ALP)
B. McCune   S. Karamchetty (ALP)

OPTICS, PHOTONICS, & DIRECTED ENERGY

A. Gschwendtner (LE)
L. Marquet   R. Buser (ALP)
S. Shey   D. Ponikvar (ALP)
R. Williamson

BIOTECHNOLOGY & BIOCHEMISTRY

C. Green (LE)   G. Petsko
J. Soukup (VLE)   D. Kamely (ALP)
B. Maidment

ADVANCED MANUFACTURING

J. Ling (LE)
R. Nagel
T. Schlie
S. Balint (ALP)

BASIC SCIENCES

G. MacDonald (LE)   J. Vesecky
J. Katz   R. Roth (ALP)
H. Levine

ENVIRONMENTAL & ATMOSPHERIC

G. Carrier (LE)
W. Lick
E. Mikhail
G. Morris (ALP)

PROPULSION & POWER

J. Elverum (LE)
W. Sarjeant (VLE)
W. Stephens (ALP)

ADVANCED MATERIALS

J. Venables (LE)   R. Katz
Y. Chiang   J. Lewandowski
L. Drzal   J. Lannon (ALP)

ELECTRONICS &SENSORS

W. Morrow (LE)
T. McGill
E. Stear
G. Iafrate (ALP)

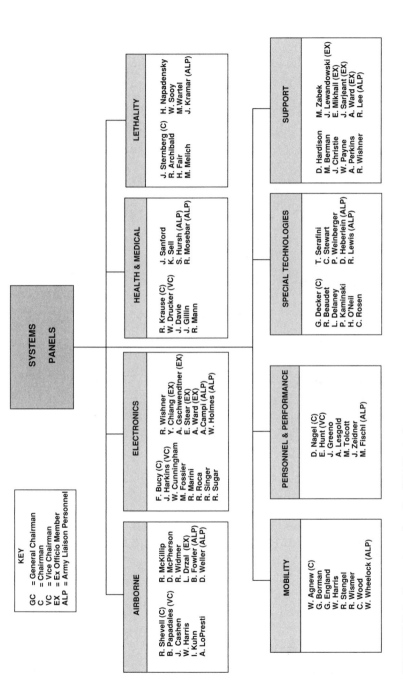

FIGURE P-1 Committee for the STAR study.

nology Management and Development Planning Subcommittee was set up. These three subcommittees reported directly to the study chairman. An Executive Committee aided the study chairman with policy guidance and served as the principal channel for communication with senior Army leadership.

The majority of the research and drafting work on the auxiliary reports was performed under the project structure described above. The Science and Technology Subcommittee and its nine science and technology groups were responsible for preparing technology forecast assessments, which are summarized in Chapter 3 and published in full as a separate volume. These assessments present the judgments of their respective science and technology groups on the likely courses of technology development over the next 10 to 20 years.

The eight systems panels under the Integration Subcommittee also prepared reports. Each of these system panel reports translates projected technological opportunities into systems capabilities that are likely to be important to the Army in the next 20 to 30 years, given the military context projected by the Technology Management and Development Planning Subcommittee. Findings from these reports are presented in Chapter 2.

The Technology Management and Development Planning Subcommittee studied the future operational environment in which the Army might find itself—potential military threats and contingencies and the missions the Army might be called upon to carry out. This subcommittee also reviewed the technology planning and management practices of the Army with the aim of suggesting improvements. Major findings from the subcommittee's report were used in Chapters 1 and 5 of the main report.

Near the end of the study, a special committee, with representatives from each of the previously constituted committees, was established to produce the main report and coordinate editorial and review work on the auxiliary reports. This committee was named the Study Committee on STAR (hereafter, the STAR Committee). Figure P-2 shows the organizational structure under which the main report and the auxiliary reports were brought to final form.

Each of the technology groups and systems panels retained responsibility for its own report. While considerable effort was made to harmonize the 19 documents, differences in substance and tone remain. The present volume, the STAR main report, reflects the consensus of the STAR Committee. Most of the other participants in the study—more than a hundred in all—agreed with most of the main report. However, complete consensus in such a large group on a

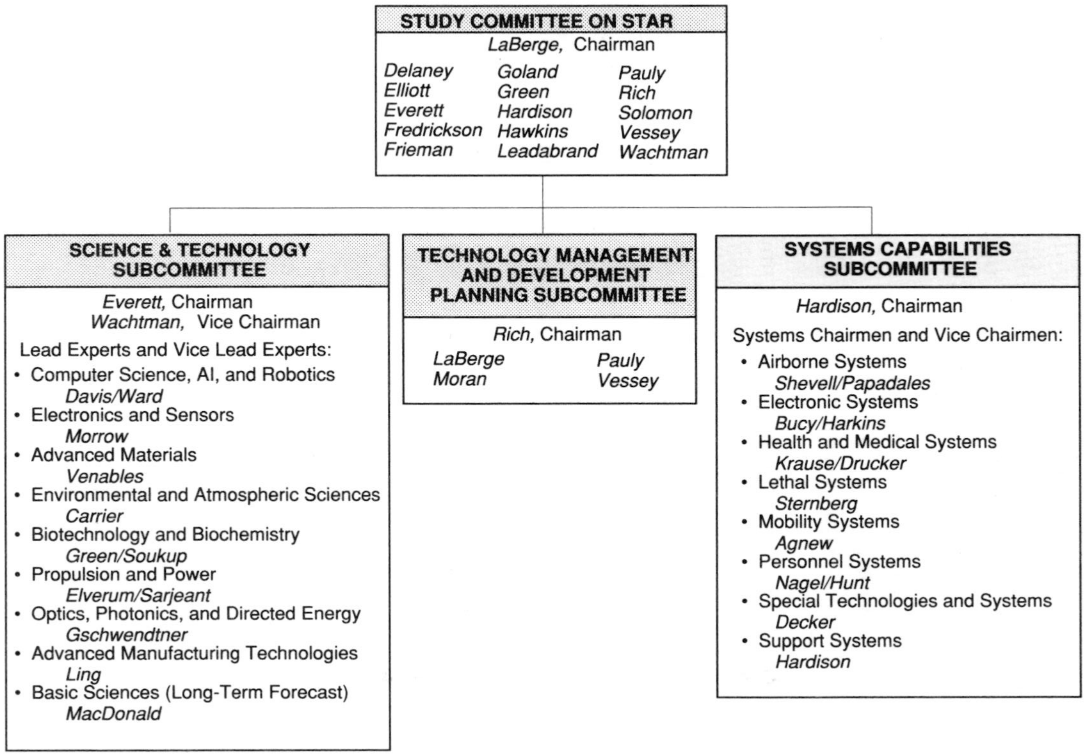

FIGURE P-2 STAR organization for final report preparation.

topic as broad as that addressed by the STAR study is impractical. Where disagreements are significant, the difference in opinion has been noted in a footnote or textual reference.

The ASA(RDA) offered the Army's cooperation in the study, to supply the technology users' perspective, provided such involvement did not compromise the independence of the National Research Council's study and review processes. High-level civilians and military officers from the Department of the Army were assigned to support the study committee. The chief scientist of the Army Materiel Command ensured that each STAR study group received support and involvement from Army personnel as desired. This was accomplished by appointing a group of senior Army liaison personnel, drawn largely from Army laboratories and procurement commands. The individual Army liaison personnel assisted the various study panels in gaining access to Army programs and activities as needed.

In addition to the frequent contact provided by the Army liaison personnel, an Army Mission Advisory Group was formed of senior Army and other service personnel, to provide a source of information about projected threats and the future environment. This group, which convened about halfway through the study, provided another means for the STAR participants to interact with Army representatives regarding the progress and appropriate focus of the study.

During the course of the study, over a hundred meetings and workshops, lasting one or two days each, were held by the various subcommittees, panels, and groups, so that members could interact with one another as well as receive briefings from the Army and other organizations as needed. In addition, three major coordination meetings were held, during which representatives of the various subcommittees, panels, and groups presented summaries of their activities for the benefit of other study participants, in an effort to identify significant gaps in coverage.

The National Academy of Sciences, the National Research Council, and the STAR Study Committee wish to acknowledge their indebtedness to the U.S. Army for its continuous and generous support and encouragement throughout the STAR study. The attention and encouragement of the top managers for Army research and development were of immense benefit. Likewise, the interest of the Army liaison personnel and the help they provided were major factors in making the study possible.

The participants also wish to express their gratitude to the STAR study staff at the National Research Council for their care and devotion to the details of arranging meetings and serving as an information center and command post while also producing and tracking an endless flow of working papers, report drafts, source materials, and correspondence.

The views, conclusions, and recommendations expressed in this report are entirely those of the STAR study members and should not be construed to represent the views of the Army or the Army liaison personnel.

The National Academy of Sciences is a private, nonprofit, self-perpetuating society of distinguished scholars engaged in scientific and engineering research, dedicated to the furtherance of science and technology and to their use for the general welfare. Upon the authority of the charter granted to it by the Congress in 1863, the Academy has a mandate that requires it to advise the federal government on scientific and technical matters. Dr. Frank Press is president of the National Academy of Sciences.

The National Academy of Engineering was established in 1964, under the charter of the National Academy of Sciences, as a parallel organization of outstanding engineers. It is autonomous in its administration and in the selection of its members, sharing with the National Academy of Sciences the responsibility for advising the federal government. The National Academy of Engineering also sponsors engineering programs aimed at meeting national needs, encourages education and research, and recognizes the superior achievements of engineers. Dr. Robert M. White is president of the National Academy of Engineering.

The Institute of Medicine was established in 1970 by the National Academy of Sciences to secure the services of eminent members of appropriate professions in the examination of policy matters pertaining to the health of the public. The Institute acts under the responsibility given to the National Academy of Sciences by its congressional charter to be an adviser to the federal government and, upon its own initiative, to identify issues of medical care, research, and education. Dr. Kenneth I. Shine is president of the Institute of Medicine.

The National Research Council was organized by the National Academy of Sciences in 1916 to associate the broad community of science and technology with the Academy's purposes of furthering knowledge and advising the federal government. Functioning in accordance with general policies determined by the Academy, the Council has become the principal operating agency of both the National Academy of Sciences and the National Academy of Engineering in providing services to the government, the public, and the scientific and engineering communities. The Council is administered jointly by both Academies and the Institute of Medicine. Dr. Frank Press and Dr. Robert M. White are chairman and vice chairman, respectively, of the National Research Council.

Table of Contents

Illustrations

FIGURES

TABLES

STAR 21

STRATEGIC TECHNOLOGIES FOR THE ARMY OF THE TWENTY-FIRST CENTURY

Executive Summary

The Strategic Technologies for the Army Report (STAR) explores the implications of new or anticipated technologies to the ways in which the U.S. Army will be prepared to fight during the next 30 years. The STAR main report is in part a summation and culmination of findings made in 18 other STAR reports, each of which focused on a smaller area within the broad scope of the study as a whole. The main report also presents the analyses and conclusions of a particular group, the Study Committee on STAR, concerning not just the subjects addressed in the detailed auxiliary reports but also certain broader issues. Finally, in addition to identifying which technology areas will most likely be important to ground warfare, it recommends a technology management strategy and projects some probable consequences of technology for the Army's force structure and strategy.

Chapter 1 introduces the report with a broad portrait of the environment the Army is likely to be facing in the next 30 years. Chapter 2 uses *systems concepts* to envision ways that the Army might use advanced technology in its principal mission areas. In Chapter 3, future prospects for individual technologies of relevance to Army applications are forecast, based on an assessment of current research and development (R&D) work in each area and expected advances. Chapter 4 describes a short list of specific technologies selected by the STAR study participants for their potentially high payoff for the Army. It also relates the various technologies forecast in Chapter 3 to key systems presented in Chapter 2. The STAR Committee's technology management recommendations are presented in Chapter 5.

Chapter 6 discusses potential implications of technology for Army force structure and strategy. Major conclusions and recommendations are summarized in Chapter 7.

INTRODUCTION: THE FUTURE ENVIRONMENT (CHAPTER 1)

How the Army uses technology in the future will be influenced by five major factors:

- an *expanding number of technology options*, as the pace of scientific and technological progress continues to accelerate;
- *changing military obligations*, as the past scenario of mid-European conflict with the Soviet Union is replaced by a broad spectrum of possible contingency operations in any region of the world, ranging from small actions like that in Grenada to major confrontations with a heavily armed army like the Persian Gulf war with Iraq;
- *diminishing funds for advanced technologies*, as shifts in national priorities and a changing world economy increase the pressure to curtail military spending;
- *closer interservice cooperation* in developing military technology and systems, in response to all three of the preceding factors; and
- *globalization of commerce*, which means the United States can no longer take for granted an unchallengeable technological advantage on the battlefield.

To respond in this environment, the Army will need the flexibility to reconfigure units rapidly for maximum effectiveness in a particular situation. The Army must be able to deploy forces rapidly anywhere in the world, while ensuring that those forces have the firepower to hold ground against an opposing force that may be larger and well armed. Real-time intelligence will be crucial to "winning the information war." Dependence on the other services and on reserves and national guard units must be planned, practiced, and coordinated so that the capabilities of deployable active Army units are enhanced rather than diminished by that dependency.

SYSTEM APPLICATIONS OF ADVANCED TECHNOLOGIES (CHAPTER 2)

Concepts for Army systems using advanced technologies are discussed under five major headings: systems to win the information war, integrated support for the soldier, systems to enhance combat power and mobility, air and ballistic missile defense, and systems for combat services support.

Systems to Win the information War

C³I/RISTA is the term used here to embrace the entire range of information-gathering functions included under the acronyms C³I (command, control, communication, and intelligence) and RISTA (reconnaissance, intelligence, surveillance, and target acquisition). In the future a highly networked system will be needed to allow integration of these functions. The sensor segment of C³I/RISTA will include large numbers of optical, infrared, radar, acoustic, and radio-intercept receivers. Robot vehicles, either airborne or ground-mobile, will become increasingly important as carriers of in-theater sensors. They will be augmented by satellite-based sensor systems and systems operated by the other services.

The communications segment of C³I/RISTA must provide quick and secure transfer of information among all the various elements in the network. Preprocessing of sensor data within "smart" sensors, wideband communications at terahertz speeds, data-compression techniques, and network management will be among the technologies needed to keep up with this communications load.

For the command and control segment of C³I/RISTA, battlefield management software will give commanders a familiar language and graphic context in which to view information, make command decisions, and have implementing orders distributed to appropriate units. Other important command-and-control aspects of a future C³I/RISTA network will be joint operability with the other services and fast, unambiguous IFFN (identification of friend, foe, or neutral) for ground systems as well as aircraft.

Integrated Support for the Soldier

The increasing technical sophistication of Army systems will not eliminate the involvement of human beings. The individual soldier will have more complex tasks to perform with more complex systems. An integrated, systems approach to meeting the needs of the individual soldier is essential. The Army's current Soldier as a System initiative is a worthwhile beginning but needs to expand to encompass the full range of soldiers' missions and the enabling technology. Within this broad sense of a "soldier system" are three areas (component systems of the larger whole) in which technology will enhance the capabilities of the soldier:

- *Combat systems* include the soldier's personal weapon and a "smart" helmet, which incorporates an audio system for communications and a visor for laser protection and built-in night vision aids. On the helmet or elsewhere, the soldier will have mission-specific options for sensors and sensor-data display devices, plus systems for navigation (mapping and positioning) and IFFN.

- *Support systems* include a personal computer (perhaps shirt-pocket size) and protection from ballistic weapons (body armor) or chemical, toxin, and biological warfare (CTBW) threats. Vaccines and bioengineered materials and medicines will protect the soldier from CTBW agents and natural disease organisms. New medical treatments and computerized knowledge bases will improve trauma care for the injured soldier both on the battlefield and during subsequent hospital care.

- *Robot helpers* will include specialized machines for hauling and lifting, airborne or ground-mobile sensor systems controlled by a single soldier or small unit, and perhaps even general-purpose systems to aid the foot soldier in carrying loads in the field and performing numerous other tasks.

Systems to Enhance Combat Power and Mobility

Long-range transport mobility will continue to rely on transport aircraft for quick deployment of light-to-medium forces and displacement ships for transport of heavy forces. To move adequate ground forces quickly to remote contingency operations, the Army must plan, design, and organize so that more of its combat power is air transportable. Sea transport will still be needed for heavy armored units to reinforce the air-deployed force and for the logistics support of deployed forces. Technology can help by allowing more systems and platforms to be air transportable, decreasing the logistics tail required to support combat operations, and improving control of materiel that is prepositioned or in the logistics pipeline.

In the battle zone, ground vehicles from transport trucks to armored fighting vehicles—including tanks or their functional equivalent—will still be used. Technological advances in the last decade have given *electric drives*, particularly in combination with advanced primary engines, more promise as propulsion systems for Army ground vehicles. Manned rotary wing aircraft (helicopters) will remain important in selected missions, although *unmanned air vehicles (UAVs)* may replace them in some roles and complement them in others. For example, helicopters probably will continue to be used for gunships and become more important in heavy-lift transport. But their scout and observation missions may soon be better performed by a range of sensor-carrying UAVs, particularly as enemy air defenses improve.

The dynamic battlefield of the future will require a highly maneuverable, armored vehicle for both assault against enemy positions and defense against opposing armor—a system with the capabilities of today's main battle tank. However, new technology will permit future tanks to be lighter and more agile without sacrificing lethal power. Stealth technology, advanced materials for armor and for signature reduction, and new propulsion concepts can maintain or increase their survivability and mobility. These new technologies could be incorporated into a tank or equivalent system designed for air transport.

The next three decades will see the evolution from today's "smart" munitions to even more "brilliant" ones, whose advanced sensors and guidance systems will allow them to be indirect-fired by artillery or rockets yet have the accuracy to destroy hard targets, including heavy armor. An advanced indirect-fire platform with multiple options for warheads is needed to give light and medium forces the capability to hold ground and interdict a much heavier and more numerous force. One warhead option is a brilliant munition able to attack moving armor; another is a less smart, high-explosive munition for attacking softer targets to an accuracy of 10 m.

Directed energy weapons that use laser or high-powered microwave beams will be available for battlefield applications. Within the time horizon of this study, they will be antisensor weapons, which are designed to destroy or temporarily blind the sensors of threat vehicles. Directed energy weapons with sufficient power to attack the hull of even light-skinned aircraft and missiles are highly unlikely to be tactical battlefield weapons within the next 30 years.

In both mine and countermine operations, new sensor technology and sensor data fusion will be key. Miniaturized sensors and processors will enable the development of smart mines: mines programmed to respond to specific target signatures and activated or deactivated remotely. In addition to distinguishing vehicle types, this technology can be used to distinguish friend from foe. For countermine operations, a number of sensor domains, including thermal imaging, high-power microwave, and laser radar are already being developed for mine detection. New techniques, such as photon backscatter, will emerge. High-power microwaves and charged-particle beams are being investigated for both detection and destruction of mines.

Robotics technology also will play an expanding role in both kinds of operations. By having the means to launch a homing projectile at a sensed target or by being mobile themselves, smart mines will have wider effective areas and the ability to attack even heavy armor successfully. On the other side, unmanned decoys that mimic the signatures of combat vehicles will "draw the fire" of hostile mines.

Air and Ballistic Missile Defense

An integrated "system of systems" will become essential for theater air and missile defense. The Army probably will not be the developer of all, or even most, of these systems, but it must be a principal architect of the system's elements and their overall integration. Ground-based target acquisition and interceptor systems will predominate, and the Army must have these elements integrated into its defensive operations. A wide range of potential threats—from tactical ballistic missiles to stealthy, low-flying aircraft, manned or unmanned, and stand-off platforms—will require a correspondingly diverse array of sensor systems and interceptors.

To overcome the inherent advantages of an attacker, these defensive systems must be coordinated into an integrated "theater airspace" defense with interoperability for all services active in that space. It must be able to distinguish friend, foe, or neutral unambiguously and sufficiently fast to allow successful interception. Many of the sensor or interceptor capabilities required of this system can evolve from current systems, fielded or in development, with the aid of anticipated technology. The integration elements for rapid detection, IFFN, target acquisition, and fire control will require new systems approaches as well as the best computing and electronics technology.

Systems for Combat Services Support

Health and medical technology developed for the military context, such as vaccines for indigenous diseases, better prosthetic devices, and artificial tissues (e.g., skin and blood), will yield benefits for civilian medicine as well. The expertise and continuing research of Army medical personnel in trauma treatment should be supported by cooperative efforts with civilian hospitals in creating one or more *trauma treatment centers*.

Other *in-theater support systems* that will benefit from new technology include (1) electronic terrain data systems; (2) improved tactical shelters based on new composite materials designed for the environment; (3) ammunition supply management systems; (4) munitions made "smarter" by advanced microelectronics and more powerful by new high explosives; (5) improved fuel supply logistics through a computerized supply tracking system, engines designed to use locally available fuel options, and better means of refueling fighting vehicles on a highly mobile battlefield; (6) reduced levels of maintenance and repair, through use of embedded diagnostics in electronic systems, more durable materials, "smart materials" with embedded sensors, and automated inventory control for parts and components; and (7) a logistics and inventory control system for Army materiel in general.

Training systems for the individual soldier and entire units will continue to advance as more powerful computers, better software, and better understanding of human-machine interactions are incorporated into Army training methods. *Simulation technology* is experiencing revolutionary advances, and the Army needs to exploit it not only for training (which it has been doing) but also for design and development, analysis of alternative tactics, and assessment of training effectiveness.

In addition to training in battlefield skills, doctrine, and simulated experience, the future Army will need personnel trained in *civic assistance* specialties. Computer-aided instruction and knowledge-base systems for cultural, linguistic, and medical information are some of the supporting technologies for these noncombat missions.

With respect to *personnel management*, the Army will be able to extend psychometric testing from its current selection role to one of classification and career counseling throughout a soldier's career. Large-scale simulation exercises can contribute to a high level of readiness, even though overseas exercises will be curtailed and specialties will increasingly be provided by reserve units.

High-Payoff System Concepts

From among the many advanced system concepts described by the STAR panels and summarized in Chapter 2, the STAR Committee selected six as having particularly high potential benefits for the Army: (1) robot vehicles (air or ground) for C^3I/RISTA missions; (2) an electronic systems architecture to provide standards and protocols for networking computers of many kinds in one large system; (3) brilliant munitions for attacking ground targets; (4) an indirect-fire system that is light enough to accompany the forces initially deployed on a contingency operation; (5) an integrated system of theater air and missile defenses; and (6) simulation systems for R&D, analysis, and training.

TECHNOLOGY ASSESSMENTS AND FORECASTS (CHAPTER 3)

The current status of technology areas relevant to Army interests was assessed by the STAR Technology Groups. Eight of these groups forecast advances likely to occur within specific technologies, in time for incorporation in fielded Army systems by 2020. There are eight corresponding Technology Forecast Assessments (TFAs). A ninth report, called the Long-Term Forecast of Research, surveys research that will open new vistas for future technology applications beyond the time horizon of the eight detailed TFAs. Major conclusions from each of these nine technology reports are presented below.

Long-Term Forecast of Research

Eleven major trends were identified as likely to draw from and have considerable influence on multiple disciplines:

• *The information explosion* on the battlefield, and in preparation for battle, will continue as intelligent sensors, unmanned systems, computer-based communications, and other information-intensive systems proliferate. Major research results are likely in third-generation data bases, mixed machine-human learning, the theory of representation creation, action-based semantics, and semantics-based information compression.

• *Computer-based simulation and visualization* will give researchers an increasingly powerful addition to traditional theory development and experimentation. Possibilities explored include a broad-spectrum physical modeling language, advanced modeling of nonlinear dynamic systems such as physical signal propagation in inhomogeneous media, and potential energy surfaces for understanding chemical reactions.

• *Control of nanoscale processes* will give the physicist, chemist, and electronics engineer the ability to create structures and devices whose dimensions are measured in nanometers, or one-trillionth of a meter.

• *Chemical synthesis by design* will allow chemicals to be designed and "engineered" at the molecular level, based on the relation between molecular structure and resulting chemical behavior.

• *A design technology for complex heterogeneous systems* could yield new ways to design complex weapons and information systems. Robustness with respect to variation will be a design objective, but nonlinear behavior in the design process itself may require a technology that focuses on the design process itself, not just the product to be designed.

• *Materials design through computational physics and chemistry* will combine the trends in computer simulation and the use of fundamental relations between structure and function to design new materials with specified properties.

• *The use of hybrid materials* will expand beyond today's structural composites to the emerging field of smart structures that react to environmental stimuli much as an organism might.

• *Advanced manufacturing and processing* will allow mass production of fine-scale materials. Nanoscale devices will be assembled into complex structures through organizing principles learned from biology, such as self-assembly and molecular recognition.

• *Principles of biomolecular structure and function* will be applied in designing new materials.

• *Principles of biological information processing* will be used to design new types of information-processing systems and to *biocouple* natural or engineered biological structures to electronic, mechanical, and photonic components.

• *Environmental protection* will affect how the Army operates and how it deals with release of hazardous materials to the environment.

Computer Science, Robotics, and Artificial Intelligence TFA

Major advances will occur in integrated system development, knowledge representation and special-purpose languages (such as battle management language), network management of diverse kinds of processors, distributed processing over multiple processors on a network, and human-machine interfaces. In these areas the Army must be prepared to invest in R&D for its requirements that do not have commercial counterparts.

Robotics will be applied to both airborne and ground-based battlefield systems. They may be fully autonomous, supervised by a human operator for nonroutine actions, or under continuous operator control (tele-operated systems). Airborne robot systems will evolve from current sensor-carrying UAVs and weapon-bearing missiles like the cruise missile. Ground-based robots will emerge as "intelligent mines" with advanced sensor capabilities, sensor data processors, and fairly simple weapons capability. They will be designed for specific missions, not as "androids" with the intelligence, skill, or versatility of a human soldier.

For the following technologies, the Army will be able to monitor and make use of advances originating in the private sector for commercial applications: machine learning and neural networks, data base management systems, ultra-high-performance serial and parallel computing, planning technology, manipulator design and control, knowledge-based systems (expert systems), and systems for processing natural language and speech.

Electronics and Sensors TFA

The three electronics technologies predicted to have the highest impact for Army applications are devices operating at terahertz (10^{12} hertz) speeds, high-speed computer architectures capable of performing 10^{12} operations per second (teraflop computers), and high-resolution imaging radar sensors. Teraflop computing will require a hundred or more processors operating in parallel at terahertz speeds. The high-resolution sensors will require both terahertz devices and teraflop computing capability.

Major advances will continue in thin-layer production methods and in expanding the number of bulk semiconducting materials used for special environments and performance higher than the current silicon-based technology. At the device level, the emerging technologies include monolithic microwave integrated circuits, superconductive electronics, vacuum micro devices, continued improvement in memory chips, application-specific integrated circuits, wafer-scale technology, microcomputer chips for digital signal processing, and better analog-to-digital converters.

At the subsystem level, data-processing applications such as signal processors and target recognizers will be implemented with multiprocessor architectures and neural networks. Smaller, more capable processors will contribute significantly to radar systems, including synthetic aperture radars, and to networks of acoustic sensor arrays.

Optics, Photonics, and Directed Energy TFA

In *optical sensor and display technology*, major advances are forecast for laser radar; multidomain sensors; sensor data fusion (performed in real time at the sensor); infrared search, track, and identification systems; focal planes designed for massively parallel data processing; and helmet-mounted or similar "heads-up" display techniques. In *photonics* (the use of light photons to transmit, store, or process information) and electro-optics (the combined use of electronic and photonic devices), the important technologies will include fiber optics, diode lasers and solid state lasers, electro-optical integrated circuits, optical neural networks, and acousto-optics for signal processing and high-speed information processing.

Directed energy devices generate highly concentrated radiation to be beamed at a small target area. The radiation used may be at optical wavelengths (as in lasers), radio frequencies (e.g., microwave beams), or other regions of the electromagnetic spectrum.

Biotechnology and Biochemistry TFA

The successes of biotechnology to date have been in medicine, agriculture, and bioproduction of specialty natural chemicals. Applications that could be developed and fielded within the STAR time horizon include deployable bioproduction of military supplies, biosensor systems, enhanced immunocompetence (resistance to disease and many CTBW agents) for personnel, novel materials with design-specified properties, battlefield diagnostic and therapeutic systems, performance-enhancing compounds, and bionic systems.

Gene technologies are methods to modify the genetic material inside cells. As knowledge of specific genes and their interactions increases, the techniques of recombinant DNA, cell fusion, and gene splicing will enable the transfer of multigene complex characteristics into cells and organisms. New substances and organisms with new properties will be produced, such as substances for discrete recognition of a particular organism or substance, compounds that modify biological responses, artificial body fluids and prosthetic materials, new foods, and organisms for decontamination.

Biomolecular engineering will use knowledge of molecular structure to create novel materials with specified properties and functions. *Bioproduction technology* uses living cells to manufacture products in usable quantities. The methods can range from fermentation, which has long been used, to multistage bioreactors. *Targeted delivery systems* are composites of biomolecules that have been structured to deliver an active chemical or biological agent to a specific site in the body before releasing it from the composite. They will be used for drug and vaccine delivery systems, special foods and diet supplements, decontamination, and regenerating or replacing tissues and organs. *Biocoupling* will link biomolecules or combinations of them to electronic, photonic, or mechanical systems. The discrete-recognition molecules developed through gene technology will have to be biocoupled to such devices to be useful as biosensor systems. *Bionics* is the technology for emulating the functioning of a living system with engineered materials. It will progress from current successes in imitating a specific biological material to eventual creation of complex, cybernetic systems that emulate the neural systems of animal behavior.

Biotechnology offers advantages over more traditional engineering and manufacturing methods for creating extremely complex substances in pure form and for very compact systems engineered at the molecular level. Exploiting the potential of biotechnology for applications specific to the Army will require multidisciplinary research teams with competence in physics, chemistry, biology, medicine, and engineering.

Advanced Materials TFA

In materials technology, three pervasive trends are forecast: (1) use of supercomputers to design materials and model performance; (2) technology demonstrators to hasten transfer of new materials and methods from laboratory to production; and (3) materials and structures designed to serve multiple purposes, thereby replacing multiple layers of single-purpose materials.

Five materials technologies were identified for special consideration by the Army: affordable resin matrix composites, reaction-formed structural ceramics, light metal alloys and intermetallics, metal matrix composites, and energetic materials. These technologies are forecast to substantially alter the state of the art for many Army applications, including armor materials, ballistic protection for the individual soldier, and weight-strength relations for vehicle and propulsion system structural design.

Resin matrix composites are becoming less expensive because of recent processing breakthroughs. The use of ordered polymers for the matrix yields composites with improved mechanical properties. Further research in molecular engineering of polymers and in matrix composition may yield organic composites with the toughness of metals and stability at high temperatures.

Smart composites have sensing elements embedded in the material. Passive sensors allow the internal properties of the material to be monitored during manufacturing and later during the material's useful life. Active elements can alter properties of the composite.

Reaction-formed ceramics can be preformed to near the final shape of a structure. Techniques for reaction-forming are forecast to replace conventional sintering technology, first for specialty components and later for even commonly used, low-cost items. Other ceramic technologies that are advancing include cellular ceramics (with foamlike structures), fiber-reinforced ceramics, and thin-film coatings of diamond or diamondlike materials.

Although some aspects of *metals technology* are considered mature, research into structure-property relations will yield evolutionary improvements even in ferrous metals technology. New aluminum alloys (such as Weldalite) and new processing techniques (such as powder metallurgy for rapidly solidified alloys) have opened up avenues for future exploration. *Metal matrix* composites are being developed that use either steel or aluminum as the matrix metal. Addition of particulates or whiskers of other metals or ceramics gives these composites the beneficial characteristics of both the matrix and the added material.

Research on *energetic materials* for Army propellants and high explosives is focusing on organic cage molecules. Another promising area of research concerns methods to make explosives less sensitive to fire, shock, impact, etc., without sacrificing explosive power. Biotechnology may prove important in the production of energetic materials and in the biodegradation of hazardous waste products from their manufacture.

Propulsion and Power TFA

In the area of *high-power directed energy*, five technologies were selected for their high potential in Army applications: (1) ionic solid state laser arrays; (2) coherent diode-laser arrays; (3) phase conjugation for high-energy lasers; (4) high-power millimeter-wave generators; and (5) high-powered microwave output from pulsed multiple-beam klystrons.

For *rocket propulsion*, gel propellants are the most promising new technology for Army applications, although evolutionary improvements to solid propellants will continue. For propulsion of *air-breathing missiles*, turbine engines and ducted or air-augmented rockets show the most potential. In *manned aircraft propulsion*, gas turbine engine technology is again the most significant technology, for both fixed wing and rotary wing aircraft. For unmanned air vehicles used in surveillance from high altitudes, high-power microwave transmission from a ground station is selected for special attention.

For *surface mobility*, primary power production, methods of power transmission, and mechanical subsystems were reviewed. Two general conceptual approaches to vehicle propulsion, the Integrated Propulsion System and hybrid electric propulsion, received highly favorable assessments. The recommended configuration combines an advanced diesel or gas turbine engine with all-electric or hybrid-electric power distribution.

In *projectile propulsion*, the two technologies selected for greatest potential are chemical propulsion by liquid propellants and electrically energized guns (either electrochemical thermal or electromagnetic).

Battle zone electric power includes primary power generation and technologies for energy storage and recovery. For continuous power generation, gas turbine engines offer more potential than the alternatives. Gas turbines for primary power and flywheels for storage would be combined with power conditioning units to supply the pulsed, short-duration power needed by high-power systems such as directed energy weapons. Rechargeable batteries are an alternative to flywheels for energy storage in both stationary and vehicle applications.

Advanced Manufacturing TFA

The next generation of progress in manufacturing will focus on the inclusion of information systems with the energy systems and material management systems developed previously. *Intelligent processing systems* use a control system to combine sensor technology with robotics. *Microfabrication*, which manipulates and fabricates materials at a scale measured in microns, will be complemented by *nanofabrication*, which does the same at the scale of individual atoms. *Computer-integrated manufacturing* organizes the single processes or workstations of a production facility into functionally related cells. Cells, in turn, are managed within factory centers responsible for system subassembly and assembly. The application of information systems to management across multiple production facilities is *systems management*.

These methods of manufacturing control by advanced information systems can be combined with specific process technologies, such as those described under Advanced Materials. Examples include distributed and forward production facilities, rapid response to operational requirements generated in the field, and parts copying from an existing part without the need for plans and specifications.

Environmental and Atmospheric Sciences TFA

The *terrain-related technologies* most important to the Army are a terrain data base that can be queried directly from the field and used to generate hard-copy maps at any scale; terrain sensing; and computerized real-time analysis of changing terrain conditions, which will use both the terrain data base and data from terrain sensors.

Among *weather-related technologies*, the Army will need atmospheric sensors flown into forward battlefield areas, either as airborne UAV sensors or ground sensors dropped in place. Satellite sensors will be used for remote sensing by laser and radar imaging. Although the Army can use advances in civilian-oriented weather modeling and forecasting, it is also concerned with modeling and forecasting on smaller scales.

ADVANCED TECHNOLOGIES IMPORTANT TO THE ARMY (CHAPTER 4)

The matrix shown on the next two pages is used in Chapter 4 to summarize the relevance of all the technologies covered by the Technology Forecast Assessments in Chapter 3 to the advanced system concepts discussed in Chapter 2.

Chapter 4 also identifies nine of the most important technologies, selected by the Science and Technology Subcommittee as a "short list" of special interest to the Army. These nine high-payoff technologies are:

- multidomain smart-sensor technology,
- terahertz-device electronics,
- secure wideband communications technology,
- battle-management software technology,
- solid state lasers and/or coherent diode-laser arrays,
- genetically engineered and developed materials and molecules,
- electric-drive technology,
- material formulation techniques for "designer" materials, and
- methods and technology for integrated systems design.

(See Appendix A for comparison of these high-payoff technologies and systems with other recent lists of technologies critical for defense.)

TECHNOLOGY MANAGEMENT STRATEGY (CHAPTER 5)

In response to the second part of the STAR statement of task, Chapter 5 recommends that the Army's technology management have a clear strategic focus and an implementation policy for how that focus can be achieved.

Strategic Focus for Technology Management

The Army should focus its technology development toward explicit Army system interests, as a means of exploiting advanced technologies more fully and of transferring new technologies more rap-

BATTLEFIELD FUNCTIONS

Legend:
- ● Advances Required
- ⊗ Advances Important
- ○ Advances Relevant

Battlefield Function Categories (column groups):

- **Information War:** Electro-Optical Sensors, UAV-Borne Sensor Systems, UGVs for C3I/RISTA, C3I/RISTA Network, Electronic Sys. Architect., Space-Based Systems, Combat Systems, Support Systems
- **Integrated Soldier Support:** Disease Prevent. & Treat., Battlefield Preventive Med., Battlefield Injury Treatment, Robotic Helper Systems, Vehicle Drive Systems, Road Building & Bridging
- **Combat Power & Mobility:** Helicopter & Transport, Advanced Armored Vehicle, Transport UAVs, Anti-Armor, Brilliant Munitions, Lt. Wt. Indirect Fire, Directed Energy Weapons, Mine & Counter-Mine Ops., Airborne Ground Attack, Air and Missile Defense
- **A & MD:** Health & Medical, Mapping Systems, Shelter
- **Combat Service Support:** Ammunitions Support, Fuel, Maintenance and Repair, Support C3, Simulation Systems, Training & Personnel

SYSTEMS CONCEPTS / TECHNOLOGIES (rows):

Comp. Sci., Art. Intell., Robotics
- Integrated System Develop.
- Knowledge Rep. & Languages
- Network Management
- Distributed Processing
- Human-Machine Interfaces
- Battlefield Robotics
- Technologies to Monitor

Electronics and Sensors
- Electronic Devices
- Data Processors
- Communication Systems
- Sensor Systems

Optics, Photonics, Dir. Energy
- Optical Sensor & Display Tech.
- Photonics & EO Technology
- Directed Energy Devices

Biotechnology & Biochemistry
- Gene Technologies
- Biomolecular Engineering
- Bioproduction Technologies
- Targeted Delivery Systems
- Biocoupling
- Bionics

Advanced Materials
- Resin Matrix composites
- Ceramics
- Metals
- Energetic Materials

BATTLEFIELD FUNCTIONS

Legend:
- ● Advances Required
- ⊗ Advances Important
- ○ Advances Relevant

TECHNOLOGIES / SYSTEMS CONCEPTS

Battlefield Function column groups:
- Information War
- Integrated Soldier Support
- Combat Power & Mobility
- A& MD
- Combat Service Support

System concept columns:
Electro-Optical Sensors; UAV-Borne Sensor Systems; UGVs for C3I/RISTA; C3I/RISTA Network; Electronic Sys. Architect.; Space-Based Systems; Combat Systems; Support Systems; Disease Prevent. & Treat.; Battlefield Preventive Med.; Battlefield Injury Treatment; Robotic Helper Systems; Vehicle Drive Systems; Road Building & Bridging; Helicopter & Transport UAVs; Advanced Armored Vehicle; Anti-Armor; Brilliant Munitions; Lt. Wt. Indirect Fire; Directed Energy Weapons; Mine & Counter-Mine Ops.; Airborne Ground Attack; Air and Missile Defense; Health & Medical; Mapping Systems; Shelter; Ammunitions Support; Fuel; Maintenance and Repair; Support C3; Simulation Systems; Training & Personnel

Technology rows:

Propulsion and Power
- High-Energy Lasers
- RF, Microwave, & MM Wave
- Missile Propulsion
- Air Vehicle Propulsion
- Surface Mobility Propulsion
- Gun/Tube Projectile Propulsion
- Battle Zone Electric Power

Manufacturing Technologies
- Designer Parts
- Distrib. & Forward Production
- Rapid Response to Field Reqs.
- Parts Copying

Environ. & Atmos. Science
- Terrain-Related Technologies
- Weather Modeling & Forecasting
- Weather Modification

Long-Term Forecast Trends
- Information Explosion
- Computer Simulation/Visualization
- Control of Nanoscale Processes
- Chemical Synthesis by Design
- Complex Systems Design
- Materials Design by Computation
- Hybrid Materials
- Advanced Manufact. & Process.
- Biomolecular Structure & Function
- Biological Information Processing
- Environmental Protection

FIGURE S-1 Relevance of STAR technologies to representative systems concepts.

idly to the field. These focal interests for the Army should fit within the larger defense policy architecture of the Office of the Secretary of Defense.

The statement of strategic focus recommends adoption of specific focal interests. The STAR Committee identified seven major potential benefits of new technology that occur in many kinds of systems across all the functional areas studied and that were repeatedly cited as important to the Army's future. These *focal values*, which should be among the Army's focal interests, are affordability, reliability, deployability, joint operability, reduced vulnerability of support and combat forces (stealth and counterstealth capabilities), casualty reduction, and support system cost reduction.

Other candidates for focal interests were selected from among the advanced systems concepts discussed in Chapter 2.

Implementation Policy

The STAR Committee recommends that the Army orient the predominant share of available resources toward those technologies and applications that are not receiving sufficient private sector investment to meet anticipated Army interest. Furthermore, wherever possible, the Army should increase its reliance on the private sector for technological progress and products.

Nine implementation actions are recommended as means of realizing this general policy:

• Commit to using commercial technologies, products, and production capabilities wherever they can be adapted to meet Army needs.

• Focus the Army's internal technology R&D on areas where strong private sector interest is not anticipated.

• Stimulate university research in technologies important to the Army that are not likely to receive adequate support either from the private sector or through other grant mechanisms.

• Balance technology funding between exploration of new concepts made possible by scientific advances and the specific technological applications needed for Army systems.

• Modernize the current inventory of systems, paying more attention to upgrading subsystems of fielded systems.

• Design systems to accommodate change and upgrading during the "design life" of a system.

• Seek to become the Department of Defense (DOD) lead agent for technologies of prime interest to the Army; consider taking on roles in other DOD programs as a means of ensuring DOD activity in areas of technology with broad utility to the Army.

• Revise Army procedures and practices to provide incentives for entrepreneurial small businesses to contract with the Army.

• Improve incentives for the private sector to invest in DOD-unique technologies, applications, and specialized facilities.

In addition to recommendations for a strategic focus and its implementation policy, the STAR Committee recommends changes in two specific areas: the Army's in-house R&D infrastructure and the Concept-Based Requirements System.

The Army's In-house R&D Infrastructure

• Shift, over time, from centers that focus narrowly on individual combat arms to each center having a broader *capability* orientation.

• Ensure adequate organizational support for Army *basic research*.

• Improve the *work environment* in Army laboratories in ways that demonstrate to the Army's scientists and engineers that their work is highly valued.

• Make the most of limited funds for in-house R&D by promoting *exchange of information* with industry.

• Attract talented technologists early in their careers and provide innovative *career advancement programs* to retain them.

• Where possible, use *rapid austere prototyping* as a design and development approach for both platforms and subsystems, to confirm applicability of new technology and as a means to validate or modify system requirements.

• Maintain a *worldwide technology watch* for advances in areas of science and technology with implications for both Army capabilities and potential enemy capabilities that need to be countered.

The Army's Concept-Based Requirements System (CBRS)

• *Keep the CBRS; alter the process.* The essential intent of the CBRS should be retained, but the implementation must be radically altered. Specific problems are addressed in the remaining recommendations.

• *Open up the front end.* The "concept" input to the requirements process should be opened up to technology exploration and to concepts built on notional threats and notional systems.

• *Ease up on Phase 1 specificity.* Lists of "must haves" and "wants" should be identified early in the requirement-generating process, but final specification of a requirement should be deferred until data gathered during development, simulation, or prototyping can be factored into the decision process.

• *Winnow as you go.* Abandon the presumption that any requirement accepted in Phase 1 research is destined for Phase 4 development. To encourage innovation, let Phase 1 be accessible to more players, but make increasingly stringent winnowing decisions at each subsequent phase.

• *Test, evaluate, and redesign.* Test and evaluation should be used as tools for learning from both successes and failures, with the lessons learned fed back into a dynamic design-redesign process.

• *Provide a vision from the top.* Rather than the current bottom-up process of requirement origination or the alternative of excessive micromanagement from above, a clear strategic vision to guide the CBRS process should be communicated from the top.

IMPLICATIONS FOR FORCE STRUCTURE AND STRATEGY (CHAPTER 6)

Two time frames are useful when assessing the implications of new technology for force structure and strategy. In the near term (within the 15-year period ending about 2005), factors such as geopolitical changes and domestic economic issues will be the dominant influences on force structure. After that time, new technologies will affect force structure and strategy more directly.

In the near term, technology can ameliorate negative consequences of these dominant factors and aid in the force structure transitions required to meet them. For example, to meet the demands of contingency operations in remote areas, (1) advanced computing and automated planning systems can provide rapid battle planning, logistics support for rapid deployment, and better joint operations coordination; (2) combat power of initially deployed forces can be enhanced with advanced antiarmor systems; and (3) troops can be prepared for unfamiliar terrain with digital terrain mapping and for an unfamiliar foe with computer-aided instruction.

The Army can also use technology to prepare for enemies who have "gone to school" on the Persian Gulf war. They may attempt to inflict sizable casualties on initially deployed American forces, particularly on vulnerable rear-area concentrations. The mode of attack could range from urban guerilla bombing missions, as occurred in Beirut, to the use of CTBW agents, tactical ballistic missiles, low-flying aircraft and missiles, or overwhelming force. Preparatory actions include priority implementation of the Soldier-as-a-System initiative, expanded use of human intelligence and counterintelligence measures, movement toward an integrated, interservice network for defense against theater air and missile threats, and fielding of direct-fire and indirect-fire systems usable by light forces at stand-off ranges.

The effects of expected budget reductions can be partially offset by increasing the combat power of the fewer forces remaining and providing them with better C^3I. The Army will also need to develop a plan with the other services to reduce overlapping functions, so that each can concentrate on its critical missions.

In the long term, more than 15 years out, the STAR Committee foresees the following influences of technology on force structure:

• *Superiority in information management* (winning the information war) will become even more important than it has been. The Army will need to pursue the latest technology and change its modes of information acquisition, distribution, and utilization, to make the best use of the new technology.

• A *flexible, multiple-tier force structure* will lead, in particular, to a new conception of medium forces. They must be air-deployable yet able to hold ground against opposing armor until heavy forces can be inserted. Also, there must be flexibility to reallocate forces from their peacetime organization, so that existing forces can be used optimally for a particular contingency. Light and heavy forces will continue to evolve toward greater combat power invested in fewer troops.

• *Integrated defense against the next generation of air threats* must protect U.S. rear-echelon support areas as well as forward combat forces, from both ballistic missile threats and low-observable, low-flying aircraft and cruise missile threats. The technology that opposing forces may possess, while lagging substantially behind U.S. ballistic missile or stealth technology, will require improved passive and active countermeasures in response. The Army force elements engaged in air defense will require close coordination with supporting elements of the other services.

• As *support and maintenance* requirements change with the increased use of smart weapons and with improved durability and reliability of systems and components, the force structure required for these activities will decrease. On the other hand, force elements associated with the full range of C^3I/RISTA operations are likely to increase. As the need for highly skilled technicians increases, civilian contractors are likely to fill more of the roles previously performed by Army personnel.

• *Training methods* will use computer simulation technology and networked wargame simulations to ensure the readiness of both active units and reserves. Experimental test units, similar to Navy

VX squadrons, could provide both developmental and operational evaluations of new technology. Simulation networks will allow coordinated training exercises in which widely dispersed units, such as reserve units with specialty skills, will participate with active duty units.

CONCLUSIONS AND RECOMMENDATIONS
(CHAPTER 7)

Chapter 7 draws together the major conclusions and recommendations from the first six chapters. The following summary recommendations are made to the Army:

- Maintain the current level of support for research and advanced technology (i.e., the funding under lines 6.1, 6.2, and 6.3a).
- Incorporate the STAR high-payoff technologies into the Army Technology Base Master Plan.
- Include the STAR high-payoff notional systems among the focal interests for an Army technology management strategy.
- Also include among the focal interests the values of affordability, reliability, deployability, joint operability, reduced vulnerability of U.S. combat and support forces, reduction in casualties and severity of injuries and disease among deployed forces, and support system cost reduction.
- Implement an expanded test program to evaluate technological opportunities and notional system concepts, in support of requirements specification and design.
- Evolve a "medium-force" tier by upgrading the combat capabilities of existing first-to-be-deployed light forces and substantially reducing the transport weight of heavy forces.
- Allocate the predominant share of Army technological resources to areas not likely to be well supported by the private sector for commercial development, while fostering cooperative efforts with the civilian sector to maintain talent and provide training (as in Army medical personnel serving at civilian trauma centers).
- Adopt and develop procedures, such as rapid austere prototyping, to expedite the movement of technology from the laboratory into the hands of its forces.
- Plan to meet future mobilization requirements, including surge manufacturing capacity and reconstitution of forces, in light of expected reductions in procurement and war reserve material levels.
- Lead, or participate strongly, in developing joint program plans, requirements definitions, and R&D in areas where there are opportu-

nities to improve joint operations with other services (e.g., airlift and sealift for first-deployed forces, C^3I/RISTA systems, theater air and missile defense, and close air support).

• Implement programs to ensure that the Army will continue to attract, train, and retain personnel of the highest quality in its advanced technology structure.

• Modify the Concept-Based Requirements Process to accelerate applications of advanced technology and to accommodate the inevitable evolution of requirements in the face of new technology.

1

Introduction

STAR STUDY OBJECTIVES

In March 1988, the Assistant Secretary of the Army for Research Development and Acquisition requested from the National Research Council's Board on Army Science and Technology (BAST) a study of the future importance of advanced technologies to the U.S. Army. In response to this request, BAST initiated a study under the general title of the Strategic Technologies for the Army Report, or STAR. This report by the Study Committee on STAR (the STAR Committee) and its companion volumes constitute the results of the BAST study.

The initiating request delineated three specific objectives of the study:

• Identify the advanced technologies most likely to be important to ground warfare in the twenty-first century.
• Suggest strategies for the Army to consider in developing the full potential of these technologies.
• Project, where possible, the implications of the technologies for force structure and strategy.

The study was to address a period extending at least 30 years into the future.

The STAR reports have primarily explored the influence of technology on conventional warfare in contingency situations. By agreement, the study has not addressed issues of nuclear conflict or the implications of technology for that class of warfare. However, in certain cases the delivery systems for conventional and nuclear warheads may be similar.

The three objectives in the initiating request have guided the entire STAR process as well as the organization of the resulting STAR volumes.[1] Figure 1-1 illustrates the STAR response to the first objective. More than a hundred technological specialties were identified by the nine groups of the Science and Technology Subcommittee as likely to have major advances that could be incorporated into Army systems within 30 years. The Science and Technology Subcommittee selected nine of these important future technologies for particular consideration by the Army as *high-payoff technologies.*

The eight systems panels included the entire range of important future technologies in their consideration of systems in which these technologies might be applied to Army functions. The STAR Committee has organized its discussion of the large number of advanced systems concepts envisioned by the systems panels by broad categories of Army function. Five of these functional categories have been selected by the Committee for their high impact on future Army operations and on technology applications. Chapter 2 reviews many of the systems concepts discussed by the systems panels for these five *high-impact functions.* In addition, the STAR Committee has selected six systems in which technology implementation is likely to be particularly valuable in the future to the Army. These six systems are the *high-payoff systems concepts.*

Chapter 3 summarizes the findings of the STAR technology groups for all the important future technologies. Chapter 4 presents the nine technologies selected by the STAR panels for their high payoff. It also relates a broad range of technologies discussed in Chapter 3 to key systems concepts from Chapter 2. Chapter 5 responds to the second point of the STAR request by suggesting a technology management strategy for the Army. The strategy calls for focusing on technology implementations in each of the five high-impact functions from Chapter 2.

Underlying the discussions of high-impact functions, either with respect to their system applications (Chapter 2) or technology management (Chapter 5), is a general perspective on the contexts in which the Army may be involved in ground warfare during the next 30 years. The remainder of this introduction summarizes this perspective and highlights key connections between it and the systems or

---

[1]The members of the STAR panels, subcommittees, and technology groups are listed in Appendix B. The preface describes the study process.

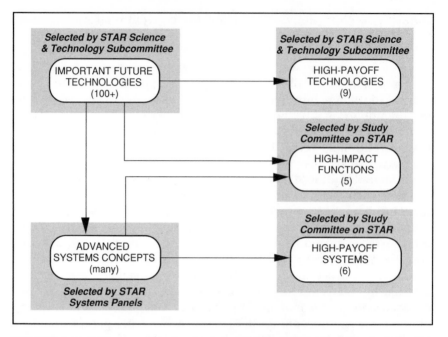

FIGURE 1-1   Selection of technologies and advanced systems concepts for STAR.

managerial implications that are elaborated on in subsequent chapters. Chapter 6 returns explicitly to this general perspective to consider some implications of advanced technology for Army force structure and strategy—the third and final objective set forth in the STAR request. Chapter 7 compiles and summarizes the major conclusions and recommendations from the body of the report.

CHALLENGES FOR THE ARMY IN THE NEXT CENTURY

The Defense Environment After Desert Storm

The U.S. armed forces recently completed one of the most successful campaigns in military history. After 40 days of air war, the 100-hour ground campaign that climaxed Operation Desert Storm liberated Kuwait by defeating a heavily armored and well-entrenched opponent. The number of coalition casualties from the ground offensive was far lower than anticipated.

While the technological superiority of U.S. forces and their coalition partners was an important factor, this success depended on capable soldiers, superb training and leadership, and a bold strategy that made the most of superior technology. Yet even as U.S. forces returned home from this triumph, major challenges lay ahead for the Army. The world remains unsettled by regional strife. The Army may again have to defend U.S. interests against a well-armed opponent under difficult circumstances. At the same time, new economic realities at home portend large reductions in force structure and in the acquisition of new armaments.

On the international scene, a still-evolving multipolar political order is replacing the bipolar world of the Cold War. Underlying this political order is a far greater economic interdependence among nations participating in an international market economy, which will become truly global within the next decade. Some of the defense-related implications of this political and economic interdependence can already be glimpsed in the joint coalition operations and cost sharing that accompanied Desert Shield and Desert Storm.

Another highly relevant element of this new order is the continuing and burgeoning reach of technology into every facet of civilian life. The controversies and recriminations over international technology transfer to Iraq prior to Desert Storm illustrate how readily a global market in information and technology can spawn important military consequences.

Because of these many changes, the Army will find the management of its technology development and the acquisition of hardware based on this technology to be particularly challenging over the next several decades. Concepts based on the past will require critical examination; those that do not fit the new environment must be discarded.

The technology management initiatives now under way to respond to these challenging times are highly commendable. Even so, the STAR Committee believes that a major reassessment of Army technology development strategies and implementation policies will be necessary in light of the considerations advanced in the STAR reports. To aid in that reassessment, this report begins by presenting what the STAR Committee believes will be the principal influences during the 30-year period covered by this study. The Army is encouraged to not only consider the issues as presented in the following paragraphs but, more importantly, to generate its own view of the dominating issues and to develop a plan for dealing with those issues.

Principal Influences on Technology
in the Next Decades

The STAR Committee sees at least five primary influences on future technology strategies of the Army:

- an increasing number of future technology options and sources;
- changing military obligations;
- diminishing funds for advanced technology;
- requirements for closer interservice cooperation in advanced technology development; and
- the "globalization" of commerce, with the attendant development outside the United States of some leading-edge technologies important to the Army.

Number and Source of Future Technology Options

Every STAR panel predicts there will be expanded technology options available to the Army for exploitation in its systems. These options will build on the advances of the past few decades in such areas as microelectronic devices, new kinds of materials and their fabrication techniques, computer hardware and software, data storage and display techniques, medical science applications based on biotechnology and radically improved instrumentation, and understanding of social behavior and learning.

The 1970s were the decade of the simple microcircuit chip and the large central processors, or mainframes, it made possible. The 1980s were the decade of pervasive use of these earlier technologies and of the introduction of a comparatively simple laser and simple microprocessor and the application of molecular biology.

The 1990s and early 2000s will see dramatic proliferation of capabilities that were just appearing as the 1980s ended. They will also mark the introduction of new technologies just beginning to be conceived. One example, typical of many, is the arrival of new materials whose characteristics will be formulated, with the aid of computers, physical chemistry, and biophysics, specifically for the structures and functions for which they will be used. No longer will functions of materials be constrained by accidental discovery of new forms of material, because these sciences now make possible true "designer" materials. By matching this explosion of technology in the commercial world, the pace of new military capabilities can accelerate at an equally rapid pace.

This rapid change of technology with military application will occur despite the prospect of reduced military budgets for technology exploitation. This paradoxical result can happen if, as the STAR Committee anticipates, the military makes better use of commercial technology. In the restricted budget environments of the future, rapid movement from technology to implementation probably can be achieved only if (1) the Army focuses its resources on those technologies not being developed for the private sector and (2) the Army develops a close working association with the private sector in those areas where applications are similar. Future Army equipment must be designed, to the greatest degree possible, to be built from commercially available parts on commercially available tooling. Mobilization of the industrial base in times of crisis will require that the Army learn how to make fuller use of commercial production capacity. It must also make suitable, planned investments to ensure that a wartime force structure can be reconstituted quickly enough should the need arise.

The Changing Military Obligations of the Army

To the members of the STAR Committee, military operations in the Persian Gulf war, Panama, and Grenada represent remarkably well the wide gamut of rapid-response contingency operations to be expected for the next decades. In addition, of course, the Army must be prepared to expand its capability to meet the potential resurgence of an adversarial major power. It must also prepare for its role in strategic defense as that role evolves from ongoing political considerations. At the other end of the warfighting spectrum, low-intensity conflicts, guerilla warfare, and counter-insurgency operations continue as real possibilities, whether for U.S forces in advisory roles or as active combatants.

Despite restricted budgets, the Army must apply resources to develop more lethal armaments for both its initially deployed, highly mobile, lighter forces and its reinforcing heavy forces. The former must have sufficient combat power and mobility to take and hold ground in a contingency situation, with air support from the Air Force and Navy. The latter will remain a necessity during the next decades for conducting offensive operations against a large, well-armed, and well-armored adversary. To make the best use of the technology opportunities suggested by this report and the supporting STAR reports, the Army will need to construct scenarios it believes are representative of the timing and extent of the military operations that might occur.

For the Army to use these technologies effectively, the technical community must have a comprehensive understanding of the Army's objectives during the coming decades. In addition, what the Army should plan to do militarily will also depend on what the technology will enable it to do. Therefore, the STAR Committee believes that the process of defining Army system requirements can be effective only if there is close cooperation between the Army's user and technology communities. Each of these communities must recognize that both will gain by achieving synergy between "technology push" and "requirements pull," instead of ineffectual tugging in disparate directions.

One important new element of these expected operational scenarios will be a radical increase in the extent of joint operations by U.S. forces. To a considerable degree beyond that observed today, new Army requirements must reflect the benefits (probable air superiority, control of the sea, satellite resources, etc.) and the difficulties (identification of friend or foe (IFF), electronic warfare, control of forces, etc.) of major joint operations.

The shift of attention from a large central war to limited contingency operations also brings a change in acceptance of casualties, within the military itself and in political support from the country at large. The Army must arm itself and plan for fighting limited combat missions with *predictably low casualties to U.S. forces and to enemy noncombatants*. In most contingency situations, national interests will be involved, but no great galvanizing principle will be at stake. Projections of casualties will be, as they recently have been, a major factor in the political decision to commit the military to warfare and in the decisions by military leaders on how to prosecute an operation once committed. By implication, technological developments that reduce the risk of casualties in either category—minimization of U.S. military casualties or of an opponents' noncombatant casualties—are of substantial value for that reason alone. More accurate operational intelligence, precision weapons, stand-off weapons platforms, protection of vulnerable rear-echelon areas, survivability of manned systems, and better treatment of the wounded are some of the more obvious areas in which this ubiquitous concern plays a role.

Diminishing Funds for Advanced Technology

The impact to the Army of constrained future financial resources cannot be overemphasized. The costs of new equipment, intended to replace fielded equipment now becoming obsolete, have skyrocketed. Even if future acquisition budgets were increasing, none of the services, including the Army, could afford to pursue all the possibilities opened by research.

For two reasons, funds for technological research and development (R&D) will decrease even if the budget lines for Army R&D remain at their current levels. First, the federal budget lines for R&D are approximately equal to the amount invested by the defense industry for in-house R&D or industry-sponsored university research. These industry investments are directly proportional to the level of military production. Because military-related production is declining, the industry contribution to R&D will decline. Second, production work on new platforms and systems contributes to a significant amount of technological R&D as the system is moved from feasibility to demonstration and eventual production. As the introduction rate of new systems declines, this source of technology funding will decline.

Both of these indirect forms of technology funding—industry R&D and new systems introduction—are much more application-oriented than the budget line for technology research. Barring unexpected reversals in the world military climate, they appear certain to continue declining from the levels of the late 1980s. This line of reasoning argues for continuation of the current level of direct federal funding for technology research to ameliorate the effects of cuts in the indirect sources of funding. Also, it suggests a clearer focus for the remaining R&D on applicability to Army-specific needs that are unlikely to be met otherwise. The latter point will be elaborated on in Chapter 5 through a suggested *implementation policy* and a *technology management strategy*.

In addition, the Army is starting from a smaller base in its acquisition funding, compared with the other services (Figure 1-2). In the Department of Defense's (DOD) accounts for research, development, and acquisition for fiscal year 1992 (FY92), the Navy portion is 35 percent of the total and the Air Force portion is 39 percent; the Army receives only 14 percent. From fiscal year 1985 to 1992, the Army's total obligation authority for research, development, and acquisition, measured in FY92 dollars, declined by 46 percent.

The explicit use of technology to achieve cost containment (and thus be able to field more equipment per scarce dollar) may be one of the most important considerations for Army technology management. The following six applications of this focus on affordability appear straightforward and will be discussed further in Chapter 5:

• Emphasize low-failure electronic, electromechanical, and mechanical design practice to reduce materiel and personnel support costs in

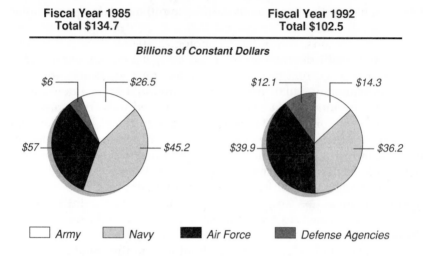

FIGURE 1-2 Changes in DOD total obligation authority for research, development, and acquisition, 1985-1992. SOURCE: U.S. Army; Office of Assistant Secretary (Research, Development, and Acquisition).

the field. This approach should be applied to complex support equipment, such as automated test equipment, as well as to weapon systems.

• Increase the use of commercial practices to procure equipment to be fielded. The success of the Army experiment for the procurement of global positioning terminals, which was undertaken before the Mideast deployment, was unequivocally demonstrated by operations in Desert Storm. That program can set a pattern for much-increased use of commercial design requirements and procurement practices. It appears to the STAR Committee that a far broader range of possibilities exists for the Army to use commercial implementations of technology.

• Plan for fuller use of commercially available capability in an emergency. An example is increased dependence on the rapidly expanding U.S. commercial air carrier fleets for rapid transport of immediately deployable forces and their equipment. The use of the Civil Reserve Air Fleet during Desert Shield operations indicates but does not fully realize this potential. Were new systems to be more oriented toward air transport in wartime, the STAR Committee believes Army capability could be greatly increased at a substantially decreased cost. The Army is already well started on this approach in its procurement of support vehicles.

• Stimulate, through economic incentives, industry investment in flexible manufacturing equipment that can be used to produce defense items at low and fluctuating rates of manufacture. Once flexible manufacturing techniques and technology have been established through these incentives, they can be applied to profitable commercial production as well as military production.

• Design platforms and equipment to accommodate change. The lifetimes of fielded designs, before they are replaced by the next-generation design, will most likely continue to lengthen. The designs of major platforms, for example, must allow retrofit with newer, more advanced components and subsystems, rather than delaying all improvement until the next-generation platform is fielded.

• To augment the Army's own funding for technology R&D, seek sponsorship from the Strategic Defense Initiative Organization (SDIO) for R&D in areas where SDIO and Army needs overlap. In particular, the SDIO offers an opportunity to ensure adequate funding for Army defenses against future tactical ballistic missile threats (Figure 1-3).

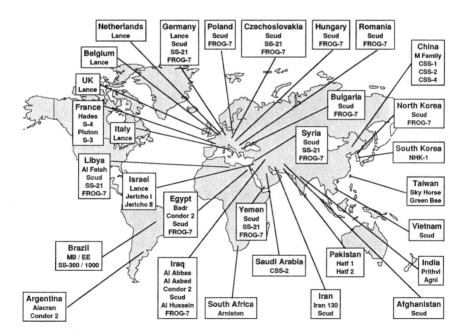

FIGURE 1-3 Tactical ballistic missile technology has already spread around the globe. (Courtesy Public Affairs Office, Strategic Defense Initiative Organization.)

The STAR Committee believes that, through whatever means the Army believes best, cost containment of technology-based systems will be of great importance. The Army should accept the challenge to find the most appropriate ways to harness technology toward this end.

Technology Program Consolidation

Duplication of effort in the technology R&D programs of the different services can be reduced by careful consolidation. The Army has already taken commendable steps in this direction through Project Reliance, which aims at achieving closer integration of these technology R&D programs. Project Reliance implements a greater degree of interservice dependence when requirements of the various services are at all similar.

The less coordinated efforts by the services to exploit directed energy for antisensor weapons provide an excellent current example of a case in which all the services would benefit from consolidated technology development and coherent central direction.

The STAR Committee believes that both the Congress and the Office of the Secretary of Defense will continue to press for more efficient use of the limited funds available to support the military technology base. Consolidation of technology programs will find many advocates. The Committee encourages the Army to continue its leadership in this area. First, the objectives cited by these higher levels are valid on their own merit. Second, there remains a considerable chance that more radical but less efficient alternatives could be forced on the services if they appear unwilling to pursue cooperative consolidation.

Globalization of Commerce

Most industrial economists acknowledge a rapid and inexorable process that is forcing the major industrial companies of the world to become global in their operations. To survive, all major industries (except perhaps those with only defense clients) will have development and manufacturing, as well as sales operations, distributed throughout the world. Most large corporate managements are well on the way to this diversification at all levels.

Global diversification will result in widespread sharing of research, development, and production technology among the internationally based elements of these diversified corporations. In circumstances where the U.S. military must rely heavily on the private sector's R&D and production infrastructure, the United States can no longer assume, as it has in the past, a substantial domestic technological leadership.

The current technological superiority of U.S. forces, which was demonstrated to the world in Desert Storm, derived in the past from two sources: (1) the superiority of American university-based basic and applied research and (2) the Army's ability to move the technology produced by research into the field. In important areas of advanced technology, U.S. universities and private sector laboratories now share leadership with research institutions in other countries. Also, as other countries continue to match or exceed the United States in R&D prowess, an increasing number of the postgraduate students in scientific and engineering programs in the United States are foreign nationals. Many of them will eventually return to their native countries. One consequence of these changes is that the United States can no longer rely on embargo on advanced technologies to provide a breathing space before other nations have access to those technologies.

Nevertheless, the STAR Committee believes that the Army can sustain its technological edge provided it can accelerate the introduction of new technology into its fielded systems. Among the means to do this is designing major platforms for change, so that subsystem upgrades move technological advances rapidly into the installed base of fielded capability.

It is important that the Army consider the major changes now occurring in the private sector that supports the Army. Many suggestions on this topic are made by the STAR Committee in later chapters and in the supporting STAR reports. However, it is up to the Army to set its own sights on the future and, in particular, to have a program that responds to the globalization of the technology base on which it depends.

CHARACTERISTICS OF THE THREATS

To prepare for its task, the STAR Technology Management and Development Planning Subcommittee began by evaluating the likely circumstances in which the Army might need to use its technology during the coming decades. The subcommittee sought to leaven the considerable knowledge and experience of its members with the insights of a distinguished group of senior retired military officers. Two special symposia, held in November and December 1989, brought these special guests together with the STAR subcommittee. The conclusions from these symposia are abstracted here to provide the context of external factors—the factors apart from technological considerations—that were used by the STAR Committee in forming its judgments, suggestions, and recommendations, which are presented in the remainder of this report.

- The dissolution of the Warsaw Pact as a credible opposition force is permanent. Conditions inside the former Soviet Union greatly reduce the probability of a conventional NATO confrontation with it.
- Although the threat has receded, Soviet nuclear and conventional capabilities continue to be substantial. High levels of intelligence gathering and verification of arms control agreements are still required. Also, increased turmoil between nations elsewhere in the world will require a geographic broadening of the intelligence program.
- Future contingencies are likely to be regional, although they may occur anywhere on the globe. The scale, terrain, climate, indigenous culture, and character of the opposing force could vary widely from case to case (Figure 1-4). The plausible potential threats during the next three decades appear to be so varied that planning should not be based on the selection of one or two specific scenarios. Force structure

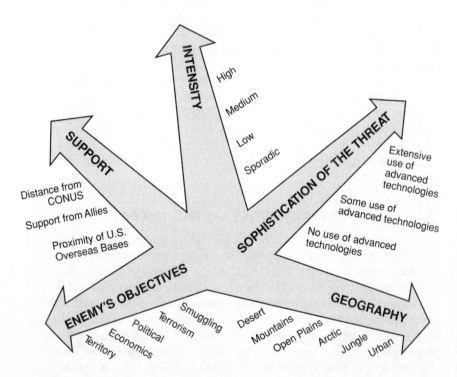

FIGURE 1-4 Spectrum of potential contingency operations.

planning that is based on a few specific scenarios could work against the flexibility needed to handle the range of situations the Army may face.

• Some of the land forces that might oppose U.S. interests are both large and capable. In both the Middle East and the Far East, potential adversaries have large forces armed with modern weapons. Some of these forces could immediately threaten areas of national interest to the United States.

• The potential for contingencies to emerge rapidly, combined with the remote locations in which the Army might have to confront them, makes timeliness of U.S. military response a far different problem than it was when reinforcement of NATO forces in mid-Europe was the dominant scenario.

• Forward basing of U.S. troops and logistic support will very likely be severely reduced, for reasons of both domestic economics and international politics.

• The ability to deploy ground forces will remain an essential element in guaranteeing credible projection of U.S. power and deterrence. Although the United States can reasonably expect to have naval and air superiority during the next decade, ground troop numerical superiority cannot be assumed in the initial stages of a U.S. military response. This situation increases the need for joint doctrinal planning and interoperable weapon development requirements.

• Prospects are encouraging for important new arms agreements among the major industrialized nations on chemical, biological, and nuclear weapons. There is also likely to be renewed emphasis on the Geneva protocols for treatment of both civilians and prisoners of war. However, the regional conflicts of the next 30 years may well involve participants who threaten to, or in fact do, ignore these agreements. The threat or actual use of chemical or biological weapons, or terrorist actions such as hostage-taking and attacks on civilians, may be used by adversaries to constrain U.S. response. These nonconventional threats require imaginative technical planning for appropriate countermeasures. "In-kind" retaliation will remain inappropriate for a number of reasons, but strategies to neutralize such threats will be needed.

• The operational performance of combat units and individual soldiers will become even more important in a world situation where both sides have access to an international market in advanced weapons technology. When the weapons on both sides are similar in technical quality, how we use our weapons will decide the outcome. In this regard, training and doctrine development will be even more important to success on the future battlefield.

CHARACTERISTICS THE ARMY WILL NEED

At these same symposia, the STAR panelists and their guest participants also considered which characteristics would be most important to the Army in meeting the threat circumstances outlined above.

- *Flexibility.* The Army must be able to reconfigure itself rapidly and on demand into operating elements with maximum deployed combat effectiveness for the range of potential threats. Mobilization of reserves, and even reconstitution of a force structure to fight a major war, should be planned along a continuum of response that begins at small-scale contingencies (such as Grenada) to be met with active units, reconfigured as needed.
- *Mobile, Survivable Combat Power.* Moving available U.S. forces rapidly to the point of desired force application becomes even more imperative and difficult under these anticipated conditions. Because the Army must be prepared to counter heavily armored adversaries, the remoteness and lack of road infrastructure in likely threat locations place a premium on firepower, especially survivable counter-armor firepower, that is transportable by air. Resupply into remote combat areas is another required characteristic. The need for rapid long-distance deployability will require designing systems to fit available air transport and, in the longer term, active Army participation in specifying new transport capabilities. In many situations there will be two aspects of mobility: (1) transport of personnel and materiel into theater and (2) battle zone mobility within theater. The range of firepower required will depend on the particular situation, but many contingency operations will require the more robust, heavier elements of the force, including tanks, armored infantry vehicles, and self-propelled howitzers, to ensure that the deployed force can accomplish its objectives while minimizing its casualties. Survivable heavy systems that can be inserted with the lighter force elements will be essential. Technology will provide the means (lighter but stronger composites, microelectronics, etc.) to make weapon systems smaller and lighter than current systems, yet at least as survivable and lethal. The ability to transport by air a small but potent force equipped with these systems will reduce the vulnerability of the "first-to-arrive" forces in a contingency operation.
- *Dependence on Reserve and National Guard Forces.* Decreasing defense budgets probably will require even further reliance on the National Guard and the Reserves for first-line Army capability. It follows that this reserve capability must be rapidly available with little advance warning.

• *Joint Operations.* Closely integrated and continuous joint combat operations will be far more important in contingency operations in the future than were required in past scenarios for mid-European contingencies. An increased emphasis on real-time combat interdependence will probably require modifications in many areas, including joint command, control, communication, and intelligence (C^3I); weapon systems design; logistics; and fighting doctrine.

• *Prepositioning of Forces and Supplies.* Limitations on transport capacity will require reliance on prepositioning forces and, more importantly, supplies. If political circumstances do not allow remote prepositioning at ground sites, prepositioned provisioning can be by ship. Shipboard provisioning may be maintained either overseas or in U.S. ports, if it is kept ready for immediate deployment. A major goal of new technology should be to reduce the currently high level of consumables expended in contingency operations.

• *Increased Real-Time Intelligence Capability.* Future intelligence requirements will be geographically dispersed beyond their current focus on Europe and the Far East. Part of the slack can be taken up by technology. Human intelligence (HUMINT) and signal intelligence (SIGINT), which will be even more important for the contingency and peace-keeping operations of the future, will be more difficult to implement effectively. For example, language skills will be needed that differ from those the Army now possesses in quantity, and real-time distribution of intelligence information must be improved.

• *Improved All-Weather, Day-Night Capability.* Among the most important advantages to U.S. forces will be the ability to operate effectively in all types of weather and continuously throughout the day and night. In addition to the requisite technology, this capability will depend on appropriate emphasis in doctrine and on training at both the individual and unit levels.

• *Psychological Operations.* As a way to reduce combat and collateral casualties, greater understanding of techniques for psychological operations (PSYOPS) and their effectiveness will be needed. The force structure and equipment to support PSYOPS must be provided.

• *Improved Short Turnaround Cycle for Planning, Deployment, and Training.* Because the nature and location of future contingencies will generally not be known with long lead times, the Army's ability to plan and train for such contingencies is necessarily restricted. Advanced simulators, knowledge bases, and other computer technology offer potential for overcoming these restrictions, at least partially.

• *Stability of the Military Institution.* The symposiasts expressed concern that a precipitate downsizing of forces might undermine the Army's capacity in the short term to respond effectively to contingency operations like those in Panama, Grenada, or the Persian Gulf. Their concern for the longer term was that the professional structure needed for a successful national mobilization, should one be necessary, might be lacking. The STAR Committee believes that technology can ameliorate these problems to some extent. Examples include wider use of computer-assisted instruction and simulation systems for training, better personnel selection and classification technology, decision-support technology to aid planners and strategists, and technology that reduces the support and logistics requirements for a given level of combat operation.

The STAR Committee emphasizes that the above descriptions of threat characteristics and of requirements for the Army to be able to meet them are not original to the STAR study. Rather, they summarize the context within which the Committee has interpreted, assessed, and integrated the findings and recommendations of the individual STAR panels. It has been the specific responsibility of the STAR Committee to formulate its report and make its final suggestions to the Army with this context clearly in mind.

A STAR VISION OF THE FUTURE

The three decades from 1960 to 1990 were undoubtedly a time of astounding technological opportunities. Furthermore, these opportunities were seized and brought to fruition in advanced Army systems. Those members of the STAR Committee whose working careers extend back 30 years and more remember the military technology of the early 1960s.

The Sidewinders and Chaparrals of that era were the technological wonders of their time. Yet they had only a half-dozen vacuum tubes to accomplish all their missile guidance. As a result, they had little capability against maneuvering aircraft or countermeasures. From our present vantage point, their military technology seems primitive. Today's Sidewinder, Chaparral, and their offspring, Stinger, have more megaflop microprocessors in them than their antecedents had vacuum tubes. These modern missiles can acquire and hit the most maneuverable aircraft under a wide range of conditions.

During this same three-decade period, today's concepts of air-land battle and high-speed maneuver became possible only by inserting new technology into heavy armored forces. The lightning left hook of the Army's heavy divisions in Operation Desert Storm demonstrated how speed, agility, accurate fire control at high speed, infrared target acquisition, and vastly improved armor have altered the tactics of tank warfare from the slow, cautious pace of single-target attack 30 years ago.

Both aircraft and satellites were used to gather intelligence then, but lengthy delays for analysis and interpretation separated the time of data acquisition from the time when commanders in the field could use the information. By contrast, both airborne and satellite reconnaissance in the Persian Gulf war gave commanders useful information in real time. The data stream was processed, communicated, and interpreted fast enough to provide early warning of a scud missile's trajectory and to guide the counterattacks. Now the sensor assets flying high above the fray can directly affect the course of battle far below.

These examples share more than just the practical use of technologies hardly imagined possible 30 years ago. Each modern marvel occurred through the vision of Army engineers who were granted the resources and freedom, by their technology managers, to explore the possible. The recommendations on the Army's in-house R&D infrastructure in Chapter 5 are meant to promote the continuation of similar opportunities for new generations of scientists and engineers.

STAR has been asked to forecast technology and systems over a similar span of three decades. None of the study participants doubt that technology will progress as much, if not more, during this next span as it has since 1960. Despite reduced budgets, there will be ample opportunities for similar success in expanding the possible to achieve the practical.

Yet the old-timers among us wonder whether the next generation of Army visionaries will enjoy an environment that encourages and nurtures their efforts and unleashes their creativity. The business of technology development has become much more complicated; it seems more difficult now to apply technology rapidly to the needs of forces in the field. The structure as it stands today casts doubt on whether the next generation will be able to seize the opportunities offered by technology to produce similar marvels in future Army systems. The implementation strategy, focal values, and other technology management changes recommended by the STAR Committee are offered in the hope of regaining an environment that will attract and encourage a new generation, to ensure the technological dominance of U.S arms into the twenty-first century.

2

System Applications of Advanced Technologies

INTRODUCTION

Eight systems panels were set up for the STAR study. Each panel was tasked with envisioning applications of advanced technologies to systems of importance to the future Army. The panel members were experts in their various application areas; most were drawn from industry. Each panel developed its own approach to performing the assigned task and wrote its own report.

In this chapter the STAR Committee has made use of *advanced system concepts* that were identified by the systems panels. These systems are exploratory; they were not carried to the point of even preliminary designs. Their purpose is to show what the technologies would be capable of doing and how the Army might use them. Further, the envisaged systems helped in assessing which battlefield *functions* will benefit most from anticipated new technology.

For some battlefield functions, there was considerable overlap among the various systems panels in systems concepts. To organize this overview of systems more simply, the STAR Committee has categorized the advanced systems concepts according to their principal function, independent of which systems panel(s) discussed them. Five functions with high impact on the Army's capability to conduct ground warfare were selected for review here:

- winning the information war (C³I/RISTA),
- integrated support for the soldier,
- combat power and mobility,

- air and ballistic missile defense, and
- combat services support.

In the section below for each of these functional headings, key systems concepts presented by the systems panels will be briefly noted. The remainder of each section will present the views of the STAR Committee on the significance of these systems to the Army, the prospects of advanced technology to affect functionality, and specific systems the Committee judged to have the highest payoff for the Army. From among the many systems concepts explored by the systems panels, the STAR Committee selected six as being of especially high potential benefit. These high-payoff systems, listed in Figure 2-1, are discussed further under their respective functions. No single high-payoff system was identified for the functional heading of Integrated Support for the Soldier. This reflects the central importance of the human *soldier* to the various systems and technology applications considered under this heading.

The STAR Committee also found, in its own deliberations and those of the systems panels, that systems and technologies were often evaluated for the same pervasive values, which cut across the requirements of particular systems or even of the broad functions listed above. These *focal values* are affordability, reliability, deployability,

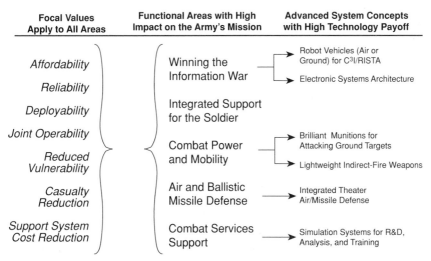

Focal Values Apply to All Areas	Functional Areas with High Impact on the Army's Mission	Advanced System Concepts with High Technology Payoff
Affordability	Winning the Information War	→ Robot Vehicles (Air or Ground) for C³I/RISTA → Electronic Systems Architecture
Reliability		
Deployability	Integrated Support for the Soldier	
Joint Operability	Combat Power and Mobility	→ Brilliant Munitions for Attacking Ground Targets
Reduced Vulnerability		→ Lightweight Indirect-Fire Weapons
Casualty Reduction	Air and Ballistic Missile Defense	→ Integrated Theater Air/Missile Defense
Support System Cost Reduction	Combat Services Support	→ Simulation Systems for R&D, Analysis, and Training

FIGURE 2-1 The STAR focal values apply across functional areas; within functional areas are advanced system concepts selected for their high-technology payoff.

joint operability, reduced vulnerability of U.S. combat and support systems (stealth and counterstealth capabilities), casualty reduction, and support system cost reduction. Their pervasiveness will be evident from the discussions below; in Chapter 5 they are addressed again as focal interests for technology management.

SYSTEMS TO WIN THE INFORMATION WAR

The current terminology for systems approaches to essential information-related functions includes C^3I (command, control, communication, and intelligence) and RISTA (reconnaissance, intelligence, surveillance, and target acquisition). Different systems panel reports use both terms without clearly distinguishing between them. This report will use the combination C^3I/RISTA to refer generally to all systems referred to by either term.

Key System to Win the Information War

The systems panels envisioned the following advanced systems concepts for the general function of winning the information war.

• *C^3I/RISTA* was addressed by the Electronics Systems Panel in three notional warfighting scenarios: large-scale operations, mid-level combat operations, and a futuristic view of how urban guerilla war might be fought. In each context the systems Panel envisioned a C^3I system that was highly robust, automated, and integrated. The component functions of the system include gathering information, evaluating and presenting this information, providing support for making command decisions based on the presentation, and distributing command decisions to implementing units. The Special Technologies Systems Panel envisioned a system of similar functions and capabilities under the heading of RISTA. The Electronics Systems Panel considered target acquisition as a separate topic.

• *Robot vehicles for C^3I/RISTA* include sensors, processors, navigation, communication, and displays, as well as the vehicles themselves, which may be either airborne or ground-mobile.

• *Electronic systems architecture* is a top-level, general information-processing architecture that will provide the standards and protocols needed to network standard serial (von Neumann) computers, signal processors, parallel processors, neural networks, and optical computers into one large "system of systems." The software provided with this systems architecture would include operating systems, communications utilities, application and user utilities, and user interfaces.

• *Space-based systems* were envisioned by the Electronics Systems Panel as tactical satellites that can be launched on demand for battlefield-specific tasks. The panel envisioned four such systems for distinct missions: communication, battle management, intelligence, and force projection. The force projection capabilities included electronic countermeasures and support measures (ECM/ESM). The basic system architecture is independent of whether the satellites are launched for battlefield-specific tasks or are joint-use satellites or national technical assets.

General Comments

To prevail in battle, the Army must gather, evaluate, and act on information more quickly than its adversary. The success of U.S. forces in the Persian Gulf war illustrates this point. The STAR Committee expects that information superiority will continue to be a key factor in future Army operations. Against a well-equipped opponent fighting with superior numbers on his home territory, it may well be the most important factor in deciding the outcome. Given the context of future threat characteristics and national policies described in Chapter 1, C^3I/RISTA requirements will expand greatly. Fortunately, the continuing revolution in hardware, software, and system architecture should provide the technology base to meet these requirements.

Today, targeting and control operations for direct-fire systems are performed in the same vehicle that carries the weapon (e.g., an attack helicopter or a tank). Each vehicle includes an internal (human) command-and-control function and therefore must support and protect the human crew. This increases the size, complexity, and cost of each vehicle. Yet in the past physical separation of the targeting and control elements from the weapon was impractical for several reasons; one of the most important was the lack of a secure and reliable command, control, and communication system. C^3I/RISTA technology and functions are likely to change markedly during the next three decades. New technologies, largely driven by commercial markets, will make possible new systems of value to the Army. The Army must remain alert to these opportunities as they emerge and, more importantly, not limit itself by rigid requirements that inhibit change.

The STAR Committee anticipates that within three decades—and possibly much earlier—all Army operations will be supported by a highly sophisticated, highly integrated C^3I/RISTA network (Figure 2-2). This network, which will supply needed information to and

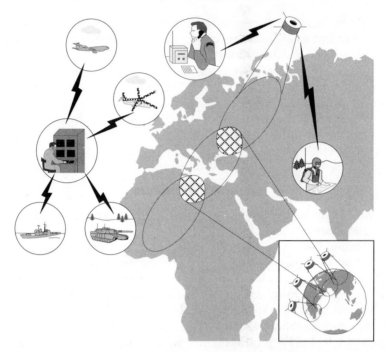

FIGURE 2-2   The future C³I/RISTA network will be both highly sophisticated and highly integrated. (Concept courtesy of Magnavox Corporation.)

from all units on and around the battlefield, will provide new capabilities to the Army. Particularly important among these will be
the capability to separate weapons physically from the system that
performs targeting.

The Army will be the principal ground force committed to the
types of operations expected to occur in the future. It should therefore play a significant role in planning for the next generation of C³I/
RISTA systems. An active Army role in this planning is critical
whether the C³I/RISTA system concept is of Army origin or the
product of a joint service effort. If a consistent Department of Defense (DOD)-wide effort does not appear, the Army should initiate—
and be willing to remain the lead agency for—such an effort.

Operational improvements can derive from both new architectures
and new technologies. Significant progress is expected in sensors,
computing and data storage, software algorithms, and communications techniques. Significant cost reductions should occur because
of broad commercial development and application of the basic
technologies.

The more detailed examination below of C^{3}I/RISTA systems divides the topic into four major functional segments, or subsystems: sensors, communications, command and control, and information management. There is also a separate discussion of space-based systems and their role in Army C^{3}I/RISTA.

The Role of Sensors

The need for information about the enemy, the terrain, and the weather is paramount in any military operation. Human intelligence aside, the means for gathering this information depend on some form of sensor positioned to receive electromagnetic radiation, sound, or other information-holding energy from the object of interest. The sensor segment of C^{3}I/RISTA includes the various sensors that perform reconnaissance, intelligence, surveillance, and target acquisition functions.

The integrated C^{3}I/RISTA systems of the future will include optical, infrared, radar, acoustic, and radio intercept receivers. They will provide comprehensive geographic coverage over a broad range of the electromagnetic spectrum. The in-theater sensor segment will be augmented by the sensors of national assets, sensors outside the theater, and sensors operated in the theater by other military organizations.

Electronic devices are the fundamental components of sensor systems. They play a role in front-end receivers and transmitters and in components for signal processing and automatic target recognition. Electronics technology for both civilian and military sensor applications is developing rapidly. Two aspects stand out as especially important for military use: the ability to form an image and the ability to respond to more than one stimulus. The former is important in identifying particular objects. The latter renders stealth by the enemy less effective.

Passive optical and infrared systems provide information on direction (bearing) and on spectral distribution and intensity; laser and radar systems provide information on reflection intensity, range, range extent, velocity, and direction. Millimeter-wave synthetic aperture radars provide high-resolution images that are responsive to the material properties of targets. These systems can be configured so that the active and passive components share the same optics and thus can provide pixel-registered images in a multidimensional space, which allows multidimensional imagery. Acoustic sensors can provide information regarding frequency and direction of detected signals. Future C^{3}I/RISTA systems will include smart processors, derived from model-based or neural network algorithms, that are able to fuse information from multiple sensor types. These smart processors for multiple stimuli also will provide the technology base for smart weapons.

An application area in which advanced sensors might achieve a major tactical breakthrough is in identification of friend, foe, or neutral (IFFN). Sensor technologies on the horizon may allow sensor systems to distinguish friend from foe without requiring a human decision-maker in the loop, thereby reducing response time and human error. Opportunities to achieve an IFFN capability by technical means alone should be pursued. As an example, if the sensors and sensor data-processing technologies forecast for "brilliant" weapons and munitions make automatic target recognition possible, these advances will not only enhance the economic effectiveness of the systems but will also contribute to solving the IFFN problem.

Sensor Placement

Depending on circumstances and the system of which the sensor is a part, a sensor's distance from the object of interest may vary from a few feet to the remoteness of space (Figure 2-3). In addition to the factor of distance, information collection with sensors requires that they be placed in appropriate positions relative to the object of interest; for example, many sensor types require an unrestricted line of

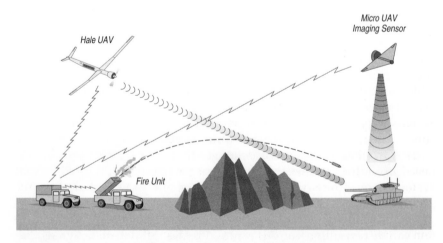

FIGURE 2-3 Remote sensor targeting capabilities in the future.

sight to the object. Of particular importance, of course, is placement that gives the ability to sense activity in areas to which soldier access is denied—for instance, behind enemy lines.

One solution is to place the sensors in a remote location, such as space, that still provides a view of the denied territory. Spaced-based systems and their potential role in Army C³I/RISTA are discussed in a later section of this chapter. Other solutions are to preposition the sensors in the area of interest or transport them there by overt or covert means. Because overt action can attract hostile reaction, methods of transporting sensors by means that either avoid detection or are relatively insensitive to enemy reaction must be sought. Unmanned air and ground systems are emerging as effective means of achieving one or both of these approaches to sensor placement.

Robot Vehicles for C³I/RISTA

Much of the C³I/RISTA information of the future Army will be obtained by satellites and high-flying aircraft using sensors that report to upper echelons, which are often located at the rear of deployed forces. After some delays and processing, selected data will flow to forward-located, small units. Besides this support, the small, forward unit will need, as it always has, highly detailed and timely information about terrain and the disposition of opposing forces. As the means to acquire such timely and high-resolution data, the STAR Committee foresees an important role for small robot vehicles operated by, and reporting directly to, the small, forward units.

The C³I/RISTA systems envisioned by the STAR systems panels include remotely controlled or robot sensors that require minimal human supervision. For several reasons, vehicles for future C³I/RISTA should be unmanned. First, in comparison with manned helicopters, which usually must also carry weapons, robot vehicles can be smaller, less expensive, more survivable, and longer enduring per mission. Second, they can acquire the needed information without pilot exposure. Command and control, managed by humans, can be performed from rear areas far from the position where the sensor performs its task. Depending on the vehicle's mode of travel, these robot sensor systems are either unmanned air vehicles (UAVs) or unmanned ground vehicles (UGVs).

Many of the battlefield sensors for the integrated C³I/RISTA system can be carried on various types of UAVs. Miniature UAVs would be deployed in large numbers. The smallest UAV might weigh no more than a few pounds and have a wing span of less than 2 ft (Figure 2-4). Each would carry a single sensor, weighing perhaps a few ounces, for periods of about a day. Deployed in groups, with each UAV carrying a different type of sensor, these vehicles would provide a robust capability for close-in C^{3}I/RISTA. Targets could be viewed from different aspects, in different portions of the electronic spectrum, and in different sensory domains.

By virtue of the large number of mini-UAVs, this C^{3}I/RISTA element would be difficult to counter by attack, jamming, or use of low-observable technology to hide ground targets. The miniature UAVs would survive because of their small size and agility in flight. Costly special treatments to give low observability would not be necessary. Costs would be minimized by using standardized airframe subsystems, produced in large quantities (thousands).

Another example envisioned by the STAR Airborne Systems Panel is an advanced form of the current high-altitude, long-endurance (HALE) aircraft (Figure 2-5). It could be extremely useful in providing continuous wide-area surveillance and bistatic illumination. It

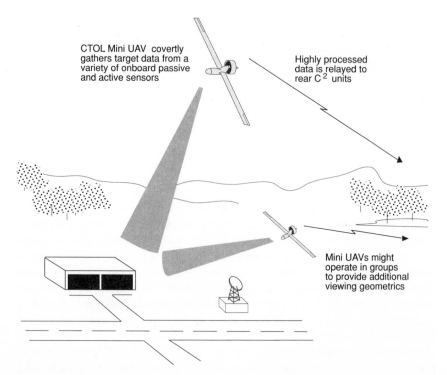

CTOL Mini UAV covertly gathers target data from a variety of onboard passive and active sensors

Highly processed data is relayed to rear C^2 units

Mini UAVs might operate in groups to provide additional viewing geometrics

FIGURE 2-4 Advanced system concept for micro UAVs.

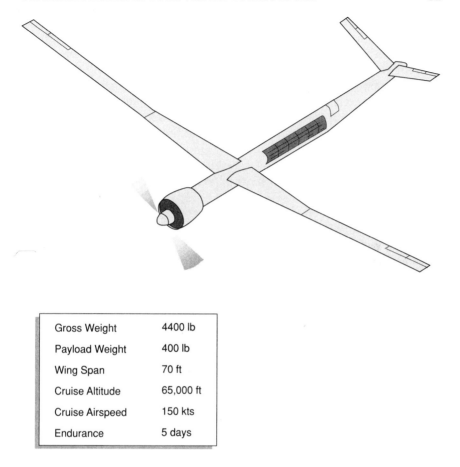

Gross Weight	4400 lb
Payload Weight	400 lb
Wing Span	70 ft
Cruise Altitude	65,000 ft
Cruise Airspeed	150 kts
Endurance	5 days

FIGURE 2-5 Advanced system concept for high-altitude long-endurance UAV.

could also act as a communication relay. The envisioned HALE UAV would weigh about 4,400 lb and carry a payload of about 400 lb; it would have a wingspan of perhaps 70 ft with a configuration typical of conventional high-performance sailplanes. The HALE aircraft could remain at an altitude of about 60,000 ft for several days. With this altitude and endurance, only a few HALE UAVs would be needed to support a typical theater of operations.

Attempts to develop UAV systems in the past have been hindered by a variety of problems, such as unreliability, complexity (requiring large, specialized operating crews), inadequate sensor and communications technology, and an inability to operate day and night under all weather conditions. The several STAR panels concerned with this application area concluded that the technologies to correct these problems are either in hand or developing rapidly enough to justify predicting the future utility of such systems as important integral elements of the C^3I/RISTA sensor segment.

Key technologies for UAV structure and propulsion include advanced composite materials; lightweight, high-endurance, high-efficiency propulsion systems; advanced fluid dynamics codes; and advanced test facilities. Other technologies are related to UAV payload, such as advanced solid state components, imaging sensors, parallel processor computer architectures, ultra-high-reliability components, signature control, and data links with high bandwidth and low probability of intercept.

Two areas in which significant progress has been made, and undoubtedly will continue, are robotics and artificial intelligence. This progress includes both improved technological performance and greater social acceptance. Robot systems will receive high-level mission orders, then will autonomously control a vehicle throughout its mission; the only human intervention may be to change mission orders. Telepresence systems may also use artificial intelligence to process voluminous sensor information and display it in ways easily understood by the remote human operator.

The UGVs envisioned by the STAR systems panels would be somewhat larger than the mini-UAV sensor systems but would have much longer mission durations (several days rather than hours). They would probably be deployed in groups. As envisioned, a UGV might weigh 4 to 20 kg and would carry a sophisticated array of sensors and processors weighing 1 to 4 kg (Figure 2-6). The sensor and processing suite would include vehicle navigation as well as C^3I/RISTA functions.

These UGVs would take advantage of their ground environment to remain undetected in enemy territory. They would use low-observable techniques for this purpose. The vehicles would require sophisticated driving and navigation systems to traverse the battlefield, remain undetected, and still perform C^3I/RISTA functions at close range.

Communications

The communications segment of C^3I/RISTA systems must provide information to whoever (or whatever) needs it, quickly and with reasonable security. Information transfer between the various C^3I/RISTA elements on the battlefield and beyond is necessary for every type and level of battle management.

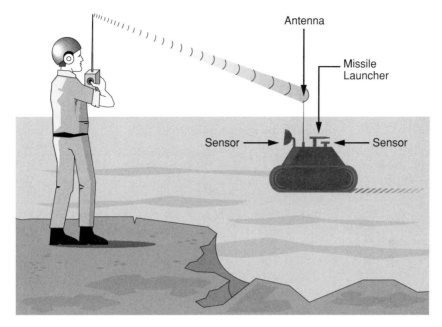

FIGURE 2-6   Tele-operated ground vehicle for surveillance, countermine operations, and so forth.

The Army currently depends heavily on a variety of land-based communication systems whose performance, support, and costs are based on technology that is several decades old and now obsolete. By contrast, the rapid progress in civilian communication systems has resulted from the use of many related technologies. These systems are increasingly global in scope. Effective yet low-cost systems for Army communications to and within remote contingency operations could be based on this commercial technology, if the mission does not involve an adversary with sophisticated signal intelligence capability.

The communications segment as envisioned by the STAR panels would use radio and optical links to connect elements of the C^3I/RISTA system in a robust network. Extreme redundancy would provide security and reliability in the face of unknown terrain, adverse weather, and enemy jamming. The network would take advantage of satellites and high-altitude, long-endurance UAVs to ensure wide-area communications connectivity.

It is critical that the Army's C^3I/RISTA system be truly robust. It should be able to support combat operations day or night and in the most adverse weather conditions. In addition, the communications network must be secure and must not experience interference from communications of other services, host nations, enemies, or neutral parties. A high degree of spectrum management will therefore be needed throughout the C^3I/RISTA system.

A successful communications segment for C^3I/RISTA will require a large carrying capacity. It must be secure, not only to prevent the enemy from determining the message content but also to prevent disruption and attempts to insert misinformation. The capacity requirement implies the availability of a wide band of frequencies. The security requirement entails physical security and the use of encryption.

The large and complex flow of data from space-based, airborne, and ground sensors will require secure, high-bandwidth links, even if data are preprocessed locally at the sensor site. Satellite millimeter and optical communications links, as well as fiber optics networks, offer the greatest potential for secure high-bandwidth transmission, for either long distances or local information distribution. Spread-spectrum electromagnetic links and fiber optic connections to remotely operated air and ground vehicles will also enable "telepresence," in which the joint capabilities of humans and machines can be optimized for many applications, including reconnaissance and targeting. The very high bandwidths provided by secure fiber optics systems will permit redundant distribution of sensor and communications information.

The advanced sensor segment of C^3I/RISTA will provide unprecedented amounts of information to be communicated, in the form of an extensive and rapid data stream. Analysis and interpretation are required before the data are useful. Very fast computers are needed for sophisticated interpretation, so computing capacity can be of critical importance. In weaponry, for example, a fast, smart missile with an imaging sensor must make complex analyses and decisions in very short times, so a high-capacity onboard computer is required. On the other hand, in many C^3I/RISTA applications, the rapid stream of sensor data must be encrypted and transmitted to high-capacity computers located elsewhere. Microelectronics—particularly terahertz devices—were singled out by the STAR panels as the heart of future ultra-fast computers. They will be needed for communication systems, data processing, and all phases of battle control. Such terahertz electronic devices will be capable of amplifying signals transmitted at frequencies of a trillion hertz (a terahertz) and switching signals at intervals measured in trillionths of a second (picoseconds).

Today's best devices approach gigahertz (a billion hertz) capability, but the STAR panels forecast a thousandfold increase to terahertz capability. The great increase in speed of terahertz devices would vastly increase communication transmission rates as well as computational power. They will have applications not only in the communication elements of C³I/RISTA systems but in many other systems as well.

Communications will be of fundamental importance to all parts of the Army's own C³I/RISTA system and to joint operability with the other services. The STAR Committee therefore recommends that the Army participate actively in the design and development of terahertz devices, which will be critical elements of future systems.

As a complement to communications technology that will carry higher data loads, data compression techniques, as well as preprocessing and fusion of sensor data at the sensor, can lighten the transmission load. For example, compression algorithms for radio transmissions, based on discrete cosine transforms, have preserved acceptable resolution and motion qualities for transmitting a television signal equivalent to 100 Mbps (megabits per second) at 19 Kbps (kilobits per second).

Command and Control

The command-and-control segment of a C³I/RISTA system is the decision-making portion. It not only performs the battle management function but also manages combat support and battlefield logistics, so that fighting forces operate in the best possible environment and are fully sustained. Another function of this segment is to manage the use of the electromagnetic spectrum, so that communications and sensors are not jammed and do not interfere with one another. Also included in the command-and-control segment will be capabilities for deception and misinformation. They will use C³I/RISTA assets to disrupt the enemy's information system or inject misinformation into it. Four key functional areas within the command-and-control segment are discussed in more detail below: battle management, IFFN, joint operability, and deception and misinformation.

Battle Management

A commander must be well informed and able to respond rapidly; this simple truth cannot be overemphasized. The enormous amount of available information is useless to a commander until it has been analyzed and summarized in a form that can be understood quickly. The commander should have just the right information and have it displayed in a familiar form.

The technology should allow a commander to call up successively more detailed levels of supporting information. The commander should also be able to test possible responses by asking "what if" questions and having a computer simulation project the expected result of an order before it is issued. Finally, the commander should be able to issue orders in a familiar form, have them translated rapidly into the detailed orders needed by field units, and have them transmitted securely to those units.

One key to any decision-making process is the ability to marshal and analyze data in a form that the decision-maker can readily comprehend. The raw capacity of computer hardware to process data has increased at a tremendous rate. The STAR Electronics and Sensors Technology Forecast predicts an order-of-magnitude increase in computer processing power during the decade to 2000 (Figure 2-7). However, the use of this capacity is constrained by the slower development of software algorithms able to dependably carry out the required types of analysis. Efficient yet reliable software is needed in the areas of intelligence extraction, synoptic organization of intelli-

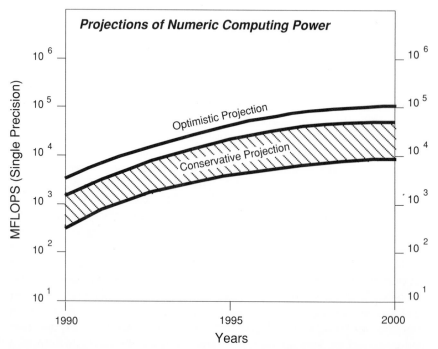

FIGURE 2-7 STAR projection for numeric computing power.

gence, and interpretation of command decisions into detailed directives to the field. In battlefield management applications, as in many other areas, software will remain the pace-setting factor.

One area of software development that does appear promising is the creation of a battlefield control language to translate command decisions into detailed directives to field units. The battle control language of the future will enable Army personnel to move data, extract information, compare courses of action, and make highly informed decisions, all without concern for computation details. It probably will be structured with layers of computer languages. The syntax and semantics of the top layer will replicate standard military operational and logistical terminology. Statements in this top-level language will look like map graphics, operation orders, or report formats.

A series of intermediate languages will provide the ability to modify software at varying levels of abstraction. As with today's spreadsheet packages, the battle control language will allow warfighting commanders to interact with the computer through a medium of commands and responses that is naturally suited to the task. The techniques of artificial intelligence can be used so that the computer understands relatively unstructured verbal commands similar to those that the user might employ in commanding human subordinates.

This battle control language might also be used as an integral part of training and analytic simulations. Its use in war games, particularly in combination with other computer simulation technology, would improve the relevance of this form of training for commanders at all levels. The same battle control software could be used to add specificity and realism to analyses, which should improve their quality.

IFFN

Another high-priority function of the command-and-control segment is IFFN (identification of friend, foe, or neutral). IFFN must be fast and unambiguous; it must not be vulnerable to exploitation by hostile forces. Moreover, consistent yet rapidly changing rules of engagement are essential.

This IFFN ability becomes even more critical in the melee of joint and combined operations, which are likely in future contingency warfare. In these new circumstances of emergency deployment, fratricide could become a major source of our casualties unless new technologies are quickly applied to this problem. Elements that will need IFFN discrimination capability include ground systems as well as the aircraft that heretofore have received attention (as in the Persian Gulf war).

In the view of the STAR Committee, battlefield IFFN is an issue to be solved technically rather than through operational approaches. As noted in the discussion of C³I/RISTA sensors, emerging technologies (such as high-speed pattern recognition as part of sensor data processing) offer new possibilities for a technical solution to this complex yet crucial problem. There are two technical approaches being studied. In direct-challenge IFFN, a dialogue between a querying system (potential defender) and the queried entity (potential threat) determines the identity of the queried entity. In a noncooperative approach, the determination of identity does not rely on any response from the entity being examined. Both technical approaches should be pursued in parallel, at least until the superiority of one approach becomes evident. A highly distributed network of sensors, extensive data bases, and sophisticated simulation systems should allow a robust IFFN capability to be developed within the command-and-control segment of the future C³I/RISTA architecture.

Joint Operability

Future deployments of U.S. forces in response to contingencies are likely to represent all the services; joint operations will be the norm. The Army will probably supply the largest contingent of forces in these operations. It must therefore focus technological assets on determining the requirements and solving the problems of command and control in joint operations. In particular, greater emphasis will be needed on managing the frequency (wavelength) and amplitude (energy) of C³I/RISTA activities than was necessary to support the warfighting scenarios for which existing Army communications equipment was designed.

A new and highly integrated joint services C³I/RISTA system will probably be a priority objective of the Joint Chiefs of Staff and the DOD in the near future. The STAR Committee believes the Army, as a matter of strategy, should aim to be a major participant in defining these joint systems and in delineating the interfaces between their components.

Deception and Misinformation

As the Persian Gulf war demonstrated, both the denial of information to the enemy and the supply of misinformation can greatly affect the outcome on the battlefield. Tactical advantage can be gained by affecting the enemy's information system; slowing the flow of, or denying, information to an enemy's information system can lessen or even negate his combat capability. A more sophisticated approach, but one having greater leverage, is to inject misinformation into the enemy's information system. Although these approaches have always been applicable to warfare, the means of implementing them have changed with the technology of military communications.

For either approach, tactical advantage can be gained by having more rapid access to relevant stored data, so that enemy capabilities and the environment are well understood. Advantage can also result from fast simulation systems, so that tactics can be quickly realigned to seize an unexpected opportunity.

As a consequence of the importance of C^3I/RISTA to potential adversaries as well as to our own side, the Army should pursue technology applications for the denial of information to our enemies and the insertion of misinformation.

Information Management

The fourth segment of the C^3I/RISTA system is information management, which includes the displays, data bases, simulation subsystems, and information processing facilities used throughout the battlefield. Local commanders need critical pieces of information quickly and in an easily understood format, but they do not need all the information available. The algorithms for information processing, filtering, and display will remain major challenges.

A revolution in performance and cost reduction is now under way in commercial information distribution. This revolution, which is forecast to continue for the next several decades, can be exploited to benefit Army information management needs. In the future, processing and simulation assets can be highly distributed, along with the data bases that support them. This will provide increased physical survivability, decreased vulnerability to enemy attempts at deception and insertion of misinformation, and increased flexibility for rapid system reconfiguration.

Space-Based Systems

Space-based systems will be extremely important to future Army operations. They will serve as enduring and nonintrusive platforms for a wide variety of sensors, for use prior to and during combat operations. They will provide a means for detecting and locating selected high-signature targets such as missile launchers. They will also support wideband communications between points that do not have surface-to-surface radio frequency lines of sight.

The Persian Gulf war saw the greatest tactical use to date of satellite systems. These systems were used for communications, RISTA, position locating, mapping, early warning, damage assessment, and environmental monitoring. As a result, the Army has begun a tactical satellite initiative. Two STAR panels (Electronics Systems and Special Technologies Systems) discussed tactical satellites as an advanced systems concept for the future battlefield. Tactical satellites are small, lightweight, low-cost systems that use advanced computer architectures and microelectronics. They can be launched in sufficient numbers to provide the redundancy needed for a robust network. Their small size greatly reduces the lift requirements for launching enough satellites to cover a remote location in a timely manner. Technology projections for the year 2020 forecast a feasible and affordable system in which a dozen of these satellites would be networked into a high-bandwidth, demand-access, packet-switched communications architecture. The system would primarily perform RISTA functions using radar optical and infrared imagery, together with signal intercept data.

The LIGHTSAT program of the Defense Advanced Research Projects Agency is a current approach to tactical communications and surveillance systems. The Army Space Command is in the process of designing a policy and program that will enable the Army to exploit the advantages of a space-based battlefield surveillance and communications system. Members of the STAR Committee familiar with the LIGHTSAT program foresee these tactical satellites as using communications protocols that will allow them to interface with existing and planned DOD satellite networks.

Tactical satellites could replace much of the force structure currently needed to perform RISTA functions, with significant savings in cost and efficiency. In addition, the coverage provided by a network of tactical satellites should be inherently well suited to contingency operations.

The extent to which the Army should own and operate the satellite systems it uses is unclear and perhaps not the central issue. What is essential is (1) that the Army have the information and communications support that space-based systems can provide, (2) that it have ground systems suited for interfacing with satellites, and (3) that the priorities for tasking space-based systems take full account of the Army's needs. Whether all three conditions can be met, absent Army ownership and operation of satellites, remains an unresolved issue, but in any case, the Army must establish its needs and consider the available options.

INTEGRATED SUPPORT FOR THE SOLDIER

Force structure reductions during the next three decades, along with a reduction in the number of skilled Americans of military age, will place a premium on increasing the capability of each U.S. soldier. In future contingency operations, initially deployed forces may be small in number and, at least at first, not fully supported by heavy forces. The effectiveness of these initially deployed troops must be enhanced by every means that technology can provide.

Even with continuing technological progress, the operation of every Army system will continue to involve a human being. No matter how sophisticated the operation of a system, human control will remain essential. Also, human interpretation of information will continue to be a major factor in system performance. Technological advances, however, will allow fewer soldiers to operate far more systems. Many tasks previously performed by soldiers will be performed by machines controlled by a single soldier, who may be located a significant distance from the machines. For all these reasons, the interface between soldier and system must be effective and efficient. Because the individual soldier will be asked to do more, and to do it with more complex systems, adequate protection and training for the individual soldier will be more important than ever.

Key Systems for Soldier Support

The systems panels envisioned the following advanced systems concepts as important to providing integrated support for the individual soldier (Figure 2-8).

- *Combat systems* include the soldier's personal weapon, nonlethal antipersonnel weapons, and antisensor weapons. They also include a smart helmet, navigation aids, cooperative IFFN, and sensory enhancement devices such as night vision binoculars and chemical, toxin, and biological warfare (CTBW) detectors. These systems concepts were considered by the Personnel Systems Panel, the Special Technologies Systems Panel, and the Electronics Systems Panel.
- *Support systems* include protective clothing (especially against CTBW agents), personal computers, medical measures against CTBW agents, rations, and special psychological training techniques. These systems concepts were considered by the Personnel Systems Panel,

Advanced Helmet

Mask

Environmental Control

Personal Weapon

Personal Computer
• Doctrine data base
• Navigation and IFFN
• Physiological monitor
• Ranging/targeting assistance

Gloves

Multi-function Clothing

Boots

FIGURE 2-8   Advanced technology can support the individual soldier in many ways.

the Health and Medical Systems Panel, the Special Technologies Systems Panel, the Support Systems Panel, and the Biotechnology and Biochemistry Technology Group.

• *Robot helper systems* include electronically controlled mechanical systems, such as an exoskeleton worn by the soldier, a "mechanical mastiff," or a "robot mule," and specialized robots for specific tasks or RISTA operations.

Training is another large area in which technology will improve the systems that support the individual soldier. Training systems are discussed below in the section on Combat Services Support Systems.

General Comments

Several STAR systems panels concluded that the most appropriate approach to developing well-integrated support for the individual soldier was to apply a systems analysis to the soldier. At one level, this approach is valuable in defining the personnel system that initially places and trains the Army's soldiers. At another level, a systems analysis can facilitate the design of the individual soldier's mission and equipment. This equipment includes the supplies, weapons, and interfaces with other systems or personnel that can enhance the individual soldier's performance.

The term "systems approach" means the organized and meticulous consideration of all component functions in a larger, operational whole (in this case, an individual soldier). Care is taken to ensure that each of the following conditions is met:

• The functions and data required for each subsystem within the larger system are available from other subsystems or from outside the system.
• All internal and external interfaces operate correctly.
• The system can function correctly in its likely environments without interference to or from other systems.

A systems approach means overall system optimization for such characteristics as reliability, preplanned product improvement (P^3I), cost, and technical risk.

With this approach to support for the soldier, two basic issues require special attention. One is the effective design of the overall system, as opposed to individual components. The other is consideration of the soldier in training and in the design stage of the equipment and systems development cycle. Accordingly, the STAR Committee believes that systems design technologies and training technology are among the advanced technologies with the highest priority for the Army.

The Army already has a Soldier-as-a-System initiative, which represents a start toward a systems approach to the individual soldier. However, this program appears to focus primarily on a particular equipment design for the foot soldier. By contrast, the STAR Committee and the STAR panels view integrated support for the soldier as more broadly applicable to soldiers with a variety of missions and not tied to a particular equipment architecture. Rather, this systems approach allows for trade-offs among options viewed as modular components or subsystems of the configurable "support system" for a given soldier with a particular mission or task to perform. This STAR systems approach is more concerned with the *continuing process* by which the soldier is equipped and protected than with any particular system *product*.

Combat Systems

Weapons

The individual soldier on the battlefields of the future must have weapons of greater lethality and range. Fortunately, weapons technology and methods of target identification offer the prospect of light, affordable armament with significantly improved capability. These may be either ballistic (bullet-shooting) or directed-energy (i.e., laser) weapons.

Future situations may also require the soldier to be equipped with weapons that temporarily incapacitate combatants or equipment. Such *soft-kill* options might include stunning combatants with "flash-bang" grenades or shells and disabling the sensors or engines of vehicles.

Navigation and IFFN

Accurate individual navigation will be possible if the soldier can exploit the previously described C^3I/RISTA network. Among other possibilities, the soldier should be able to use navigational information from satellites. The C^3I/RISTA network will provide the individual soldier with details of the location of friendly and hostile forces, terrain, and weather. The system can also include a cooperative IFFN system, which will pass individual soldier identity and location to the C^3I/RISTA network, and from there to other field and command elements.

Sensory Enhancement

In the face of CTBW threats, an essential factor in sustaining the effectiveness of combat troops in the battle area is highly dependable advance warning of the presence of a CTBW agent. This capability will require a system of sensor elements that are synergistic with one another and that have the sensory capability of the soldier who uses them.

For soldiers to perform their mission, they must detect and identify all threats and then make an appropriate response. The STAR Committee believes that sensory enhancement similar to that used in aviation will be feasible and useful for the individual combat soldier. Personal night vision and optoelectronic sensors will be useful. They may be augmented by chemical and biological detectors. Progress in microelectronics, photonics, and biotechnologies should enable a robust sensory enhancement system to be part of the future infantryman's standard equipment.

Smart Helmet

The helmet will remain an essential part of the soldier's personal equipment. It will continue to provide ballistic protection, but it can also provide an audio system for the soldier to hear communications and equipment signals. The helmet undoubtedly should include a visor for laser protection. The visor might also be used to provide holographic images from various sources, including the soldier's personal sensors.

Special helmet-mounted sensors could track the soldier's eye movement to aim personal sensors and weapons. For instance, a soldier might look at a building at a distance. A laser rangefinder and the navigation system could quickly determine the building's exact location. The soldier could provide audio information about the building through a helmet-mounted microphone. All this real-time information could be stored in the soldier's personal computer or transmitted through the $C^3I/RISTA$ network.

The helmet and visor conceivably could be used to aim the soldier's personal weapon. Current weapons depend on tight hand-eye coordination for aiming; the problem is that the eye is accurate, but the hand is not. Eye-only aiming is certainly possible with emerging technologies.

Support Systems

Particular attention should be paid to technologies that help the soldier survive on the battlefield and subsist under conditions of field deployment. Technologies that support either prevention of casualties or treatment when they occur should receive substantial program support. Among the wide range of foreseeable technological opportunities, this report will briefly review personal computers for the soldier, body armor and protective clothing, battlefield medicine, countermeasures to CTBW, and rations.

Personal Computers

Improved computer memory technologies will allow the individual soldier to carry an enormous quantity of information in a small package. A 500-megabyte digital memory in a shirt-pocket device will be possible within a decade. The issue of how to exploit this capability requires a systems approach. For example, the memory could be used to carry details of anticipated threats, medical procedures, equipment maintenance and repair, or terrain and mission. Expert systems could use this memory to analyze options and suggest alternatives.

Multifunction Clothing

The future soldier will be exposed to a wider variety of lethal threats. Against ballistic weapons, evolutionary improvements in body armor are forecast. Armor that conforms to the body and is integrated into the standard uniform appears achievable. In addition, chemical and biological protection can be built into the standard uniform. This uniform could also be made compatible with environment-control equipment for supplemental heating or cooling. Other emerging technologies may reduce the observable signatures of the soldier by reducing infrared emissivity or changing the patterns and color of a uniform while it is being worn.

The physical barrier provided by protective clothing or special gear will continue as a major element of CTBW defense. Medical intervention after heavy exposure will not completely neutralize the effects of some agents, such as simple corrosives, like phosgene, or highly potent nerve agents. The primary concern with physical protection is the degree to which personal gear degrades the soldier's task performance. This degradation is presently estimated to exceed 50 percent for some tasks, depending on ambient conditions. The causes include restricted vision, heat buildup, and impaired dexterity.

Biotechnology can play a role in improving physical protection, primarily through the development of novel materials that control the permeability of clothing to certain molecules and aerosols. In general terms, the new materials must be lighter, stronger, more selectively impervious, and cheaper than current materials, while providing sufficient heat and water vapor transfer. Novel concepts include combining the lightness and strength of silk or Kevlar-like fibers with the sheet characteristics (for imperviousness) of rubber-like compounds. Pores for heat and water vapor transfer must exclude the CTBW agents, perhaps with special chemical catalysts or enzymes embedded as "pore guards." Blast-attenuating biocomposites are already in prototype evaluation.

Biotechnology will be able to produce both natural and artificial materials, such as composites and customized polymers with specifiable physical, chemical, and electrical properties. Advances will depend on the simultaneous development of computer-aided biomolecular design and low-temperature manufacturing techniques. In 20 years, composite materials may exist that incorporate CTBW barriers, special impedance-mismatching characteristics to attenuate blast and sonic interactions, and some defense against white phosphorus munitions.

Battlefield Medicine

Even with the anticipated increases in the lethality of threat weapons, improved survivability and the deployment of fewer soldiers in close combat probably will reduce the number of casualties. Nonetheless, severe casualties can be expected. The future battlefield will be characterized by a fast tempo of operations and rapidly moving forces. Combined with highly effective antiaircraft threats, these conditions will make airborne medical evacuation far more difficult, if possible at all. The STAR panels forecast that new equipment technologies and pharmaceuticals will aid in resuscitation and trauma treatment on the battlefield. This aid will be administered by ordinary combat soldiers without requiring trained medical personnel to be present. Continued advances in the medical treatment of trauma will be a major factor in reducing fatalities and the number of incapacitating injuries. (Trauma centers for research in this area are discussed under Combat Services Support.) New materials, including those produced through biotechnology, will improve prosthetics and make possible replacement tissues such as skin and artificial blood. Expert systems for medical diagnosis, contained in hand-held computers, will allow nonspecialist personnel to make rapid and accurate diagnoses.

Medical technologies also offer promise in reducing the soldier's susceptibility to disease and to chemical and biological agents (Figure 2-9). The soldier's immune system will be enhanced for broader protection from naturally occurring infectious disease organisms, which probably will continue to be the largest cause of casualties in combat situations. Research into the mechanisms of human immunity, combined with genetic engineering and bioproduction technologies, will expand the range of vaccines and other means of enhancing the soldier's immunocompetence. Recombinant DNA technology will be used at hospitals to isolate disease organisms and produce specific vaccines within days.

Countermeasures to CTBW

The Persian Gulf war again showed how easily an adversary can use even the threat of CTBW to tactical advantage. The best response is to have ready a comprehensive array of countermeasures. The STAR Committee sees four distinct areas in which such counter-

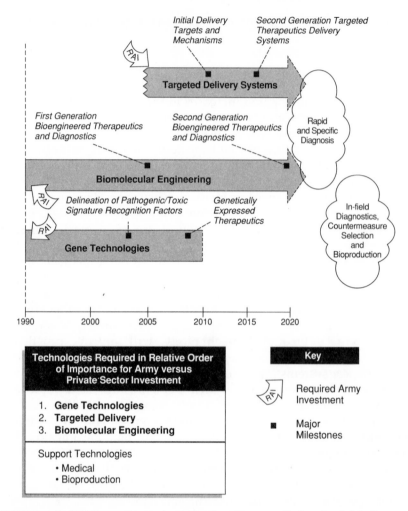

FIGURE 2-9 STAR road map for biotechnology applied to in-field diagnostics and therapy.

measures will be needed. One of these, protective clothing, has been discussed. The three discussed below are detection and identification of the CTBW agent, medical prophylaxis and therapy, and decontamination.

Detection and identification of the CTBW agents being used can be difficult, especially in the case of biological agents. Today's detection and identification techniques are based on analytical, physical, and immunological chemistry. They are too slow in detecting some agents, do not detect all agents now known, and are subject to false alarms that greatly reduce their effectiveness. Moreover, present techniques and even the next generation of mass-spectrum detectors require that we know some specific distinguishing characteristics of the agent molecule or organism far in advance of fielding effective means to detect it. These techniques will never keep pace with the evolving threat. Thus, the single most perplexing problem in detection and identification of threat agents is that we do not have, and will not have, a comprehensive list of them. This problem is particularly severe for toxins and biological agents.

The rapid development and enormous potential of biotechnology offer the best hope for fast and accurate identification and response. Embedded in the inherited information of every organism (i.e., in its genome) is highly specific information on the molecular sequences of its component biomolecules. Biotechnology can exploit this information to design and assemble biological molecules and structures that can distinguish unequivocally between agents and nonagents with similar characteristics. Gene technologies (along with the medical and biological understanding they have produced), biomolecular engineering, and biocoupling will be able to move CTBW detection and identification into the next generation of defensive strategies and beyond. Within 15 to 30 years, biosensors derived from the human immune system will provide early warning of CTBW agents.

The requirements for medical prophylaxis and treatment of CTBW agents differ somewhat from those for detection and identification in that sometimes a common approach can be used against a class of agents that operate by a similar mechanism. Nerve agents are an example of such a class. However, many of our present medical countermeasures target specific agents, especially in the area of toxins and disease-causing organisms.

To be as generic as possible, future programs must pursue such ideas as blood-borne "interceptor" molecules (for blood-borne agents), blood filtering technologies, counteragents that block the cell receptors targeted by threat agents, and targeted delivery of drugs to specific body sites. Barrier compounds applied directly to the skin are yet another direction of research. This prophylactic approach, coupled with counteragents that possess "sacrificial" binding sites for threat agents or agent-degrading moieties (enzymes, etc.), could reduce the need for physical protective clothing and gear.

The best-known current applications of biotechnology are in the areas of medical prophylaxis and therapy. Biotechnology will be essential to the success of biomedical interventions that defend against

CTBW agents that act at specific receptor sites. For example, biotechnology offers great promise for countering blood-borne toxic molecules. For prophylaxis, bioengineering possibilities include catalytic "interceptor" molecules that would mimic the agent's target site, degrade the agent molecule after it binds to the interceptor, and reset themselves for another agent molecule, all at high reaction rates. Therapeutic concepts include extracorporeal filtration, similar to kidney dialysis, but using filter beds containing antibodies to the CTBW agent. Broad-spectrum protection against pathogens will be feasible by pharmacological blockage of initial cell-binding receptors. As we learn more about how the immune system recognizes pathogens and mobilizes against them, new methods for prophylactic "exposure" and stimulation of the immune system will enhance immunocompetence.

For CTBW therapy against previously unknown agents, biotechnology offers the same kind of rapid identification and treatment potential that was discussed above for naturally occurring pathogens. Removal of agents *within* the exposed soldier (a kind of "internal decontamination") will become possible by coupling rapid identification of the agent at field hospitals with rapid antibody production systems that automate the sequencing of nucleic acids or proteins and synthesize appropriate antibodies.

For some applications, biotechnology will be combined with other advanced technologies. New methods for administration and delivery of prophylactic and therapeutic agents will use drug micro-encapsulation and targeted delivery systems. Rapid in-field diagnosis and triage for CTBW casualties by nonmedical cohorts will become feasible by combining biotechnology with medical diagnostic expert systems.

Decontamination countermeasures will be required for several categories of CTBW exposure: personnel, battle equipment, support facilities, and terrain. Most of the work on decontamination has focused on the chemical part of the CTBW threat; much less work has been done on toxins and biological agents. Current methods for decontaminating equipment have not changed greatly for decades. Decontamination of personnel relies on resins and washing, which produces contaminated waste. Decontamination of electronic equipment relies primarily on hot air to degrade chemical agents. Current decontamination procedures, such as washing with the corrosive, decontaminating DS2 solution; hot air blowers; resins; and scorched earth for decontaminating terrain, can all be improved upon.

Biotechnology offers enzymatic techniques for decontamination of personnel, for smaller surface areas, and for terrain. The terrain application would be a form of bioremediation, akin to the use of bioengineered organisms to attack oil spills at sea. Solutions of genetically engineered cells, such as macrophages, could be developed for decontamination of toxins and biological agents in circumstances where corrosive or toxic compounds cannot be applied, as in decontaminating skin or wounds.

Rations

Combat soldiers of the future probably will have lightweight, highly nutritious rations and personal means of water purification. Although the technologies will be available to produce these rations and water kits, a production base must be created to ensure the quantities that may be needed for a prolonged conflict. The STAR Committee does not foresee a large commercial market for these products.

Strength, endurance, and cognitive skills might be enhanced by using dietary supplements. In addition, safe drugs that maintain alertness for periods of 24 to 36 hours might become available. A comprehensive systems analysis of the physiological, pharmacological, and psychological implications will be needed to define the appropriate use of these opportunities.

Robot Helper Systems

As an aid for the individual soldier, the STAR Special Technologies and Systems Panel discussed a robot vehicle concept called the robot mule (Figure 2-10). The Personnel Systems Panel reviewed a functionally similar concept called a mechanical mastiff. Other concepts in this category vary from an electromechanical exoskeleton (described by the Mobility Systems Panel and Personnel Systems Panel) to simpler, specialized systems for C^3I/RISTA and for hauling, lifting, or positioning heavy ordnance and other supplies (Special Technologies Systems Panel; Personnel Systems Panel; Technology Group on Computer Science, Artificial Intelligence, and Robotics). The robot mule is discussed here as representative of the wide range of capabilities that have been envisioned for these various aids to the individual soldier.

The mule, which would carry most of the solider's load, would have a range and speed compatible with a walking soldier. It could be controlled by voice and perhaps eye movement of its soldier operator. A compact energy supply with a low heat signature will be essential and probably will be the greatest technological challenge.

The robot mule (or more specialized robots) could be designed to clear mine fields or provide short-range reconnaissance. Another

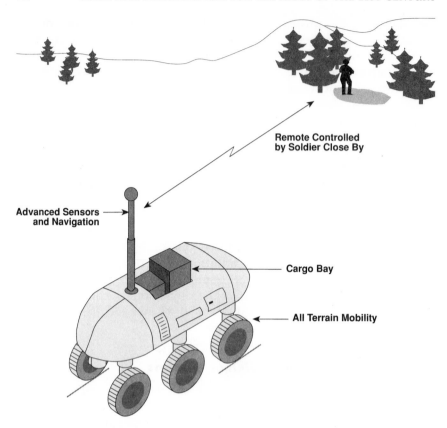

FIGURE 2-10 Concept for a robot vehicle as a soldier's aid.

significant capability would be carrying a wounded soldier to medical facilities. For either of these missions, the system would need autonomous navigation capability. The design need not incorporate a full suite of sensors to support all reconnaissance and navigation tasks in every mule; instead, separate "clip-on" sensor suites could be fitted according to the soldier's mission.

A system like this robot mule will not be as extreme an advance as it might appear. The STAR panels forecast significant progress in robotics and supporting technologies by the private sector. The Army's challenge, here as with other advanced systems concepts, will be to exploit the growing industrial technology base to fulfill its requirements.

Other robot helpers are conceived as much more specialized than the robot mule or analogous multipurpose systems. Some robot vehicles for C^3I/RISTA, such as mini-UAVs or small ground-based sensors, would be well suited for use by individual soldiers or small units. Other specialized robots might perform heavy lifting and hauling operations or dangerous operations such as minefield clearing. Multiple units could operate under the supervision of a single soldier.

SYSTEMS TO ENHANCE COMBAT POWER AND MOBILITY

In many types of likely future contingencies, the U.S. military will have little time to react if armed intervention is to succeed. Meeting this challenge will require rapidly transportable light forces, which nevertheless have sufficient combat power to defend against an opposing heavy force. Equally important will be more rapid deployment of follow-on forces and materiel with the sustained combat power to prevail against any opponent.

Key Systems for Combat Power and Mobility

This section considers advanced systems concepts in three areas central to these contingency response requirements of the future: long-range transport mobility, battle zone mobility systems, and lethality systems.

• *Long-range transport mobility* deals with the means likely to be available to move both light and heavy forces to an operations theater far away from their bases. Providing the means of such transport is not in itself an Army responsibility, so no systems concept for long-range transport is presented here. The Army should, nonetheless, specify its lift needs and pursue their fulfillment aggressively, whether they are to be met by military or commercial aircraft. Any consideration of future combat power must take into account the constraint to deliver combat systems to the theater of operation quickly and over long distances. The issues are discussed here for their relevance to advanced battlefield mobility and lethality systems concepts.

• *Battle zone mobility systems* include vehicle navigation systems and drive systems; heavy, tracked combat vehicles; light, wheeled combat vehicles; an advanced personnel carrier; individual soldier or small-unit movers; and road-building and bridging systems.

• *Lethal systems* include antiarmor weapons, brilliant munitions, an advanced indirect-fire system, directed energy weapons, and mine and countermine operations. Two of the STAR Committee's high-payoff systems concepts are in this area:

1. *Brilliant munitions for attack of ground targets.* These munitions, which will be used primarily in indirect-fire systems, will have autonomous target acquisition, hit, and kill capabilities.

2. *Lightweight indirect-fire weapons.* This system will provide a new capability, especially to light forces, for efficient indirect-fire attack, from extended range, on a broad variety of ground targets, including armored vehicles. For armored or hardened targets, brilliant munitions could be fired. For area coverage of soft targets, conventional high-explosive munitions would be fired from the same system.

General Comments

As the major weapon user of the future, the Army needs to take the lead among the military services in exploring new techniques and technologies for effective and affordable land-based weapons. To do so will require a substantial and focused program within the Army and its supporting industrial base. Equal attention must be given to affordability and performance. The STAR Committee expects that significant new challenges, and responsive concepts, will emerge as the conditions of contingency warfare are evaluated and as budget constraints force reductions in the cost of complex weapons.

Weapons undoubtedly will become more accurate, more lethal, and more expensive in the next few decades. Better weapons, which can take advantage of the improved targeting and IFFN provided by the envisioned C^3I/RISTA system, offer the potential for dramatic improvements in the Army's warfighting capability. Much of their advantage may result from coordination with physically separate RISTA systems that will allow the weapon to operate in an indirect-fire mode. For instance, an air vehicle (probably unmanned) that carries sensors will be used to locate targets. This information will be transmitted through the C^3I/RISTA network to a fire unit, which will launch the attack against the target.

In this architecture both the sensors and the weapons can be optimized for their specific functions, and full advantage can be taken of advanced technologies. The survivability of both C^3I/RISTA and weapons assets should increase. Together, these improvements portend an unprecedented flexibility in Army combat operations and a substantial enhancement of the combat power of the reduced contingent of soldiers on future battlefields.

The Importance of Long-Range Transport Mobility

The Army of the future will, without doubt, require improved long-range transportation for its forces. There is no reason to expect a breakthrough in the classic trade-off among speed, payload, and cost. Aircraft, which are relatively fast but expensive, can, realistically, transport only light forces. Conventional displacement ships are relatively inexpensive and can carry heavy forces, but they are much slower.

Carriers that use advanced technology (such as surface-effect ships or wing-in ground-effect vehicles) offer greater payload than aircraft and higher speeds than conventional displacement ships but at high developmental and operational cost. The STAR Committee concluded that the Army should expect to use conventional aircraft and displacement ships for long-range transport of its forces. The Army should pursue development of lighter gear to lessen the load; it should also take the lead in interservice planning to define carriers to transport troops and materiel.

As the first to be deployed, light forces probably will require a dedicated fleet of military transport aircraft because of the need for quick response and the possibility of hostile fire at the delivery area. These aircraft, of course, would be developed and operated by the Air Force. The Army's role will be, as it is today, to influence DOD and Air Force plans and budgets to ensure that this capability is provided.

Transport aircraft technologies are quite mature. The Army can expect substantial improvements in range and fuel economy. These advances will be driven primarily by the needs of civil aviation. Further advances, which will be driven primarily by civil aircraft needs, will not produce dramatic gains in performance, although fuel economy will improve somewhat.

Growth in the domestic fleet of long-range commercial transport aircraft makes greater use of that fleet as the Civil Reserve Air Fleet (CRAF) very attractive for movement of specific elements of both light and heavy forces. These elements include personnel, supplies, and smaller items of materiel. Returning aircraft can carry wounded, should that be necessary. The CRAF mobilization during Desert Shield was successful, and the STAR Committee expects increased use of this resource. More Army systems could be designed to fit in CRAF aircraft. Modern design techniques may allow new systems to be built into modules specifically designed for quick loading and unloading from CRAF aircraft.

The delivery of heavy forces will require long-range transportation as well, albeit slower. Displacement ships offer the only reasonable means to carry heavy forces. Modest improvements in marine technologies probably will reduce costs and decrease activation time. Sea-mobile POMCUS (Prepositioning Of Materiel Configured to Unit Sets) appears to be an attractive option for many threat scenarios. However, storing materiel on ships for extended periods will require the technology to (1) test long-stored equipment remotely, (2) reduce the support effort required, and (3) move the prepositioned materiel rapidly from storage to field use. A program that expressly addresses technology requirements for long storage life and low maintenance appears warranted.

The initial force deployment, while probably the most stressing test for long-range transport, is not the only requirement. The initial force must be reinforced, supplied, and sustained. The timing of resupply will usually be less critical than the initial insertion of forces, so much of it can be delivered by surface ship. However, because some supply needs will be time-critical, continued air transport will be required. Technology that can reduce this "logistics tail" will be of increasing value to the Army. Advanced technologies can contribute to this objective in several ways:

• increasing the effectiveness of each consumable item, such as ammunition, so that fewer units are needed;

• improving the reliability of equipment so that fewer replacement parts are required; and

• applying modern techniques for inventory management to reduce the materiel held in the logistics pipeline.

A special transport issue addressed by the STAR panels was the need for methods to insert and extract Special Operations Forces covertly. By definition, the success of such missions depends on the transport aircraft remaining undetected by the opponent. The conditions of operation also require vertical takeoff and landing (VTOL). The combination of nondetection, VTOL, and reasonable range and payload makes this an expensive aircraft to develop. Since only a few would be required (perhaps no more than 20), unit production costs would be high. These factors led the STAR Committee to agree with the systems panels that a special transport aircraft for this purpose, although technologically feasible, is probably not economically supportable. No technological advances, even as remote possibilities, are anticipated that would alter this conclusion. Special Operations Forces will therefore need to continue relying on helicopters and conventional aircraft.

Battle Zone Mobility Systems

Mobility within theater will become increasingly important in contingency land warfare that is fluid, dispersed, and fast paced. This section presents several systems concepts that apply advanced technologies to increase the capabilities of ground mobility systems. A special airborne system for battlefield mobility, the heavy-lift UAV, is discussed in the separate section on UAV systems. A mobility system geared to the individual soldier, the robot mule, was discussed above, for the soldier as a system.

Ground Vehicle Drive Systems

Advanced technologies for propulsion, drive, and traction not only will improve the performance of ground vehicles but can also lower their observability. The alternators, controls, cables, and motors of electric drives will be somewhat smaller and have less rigid space requirements than the transmissions, gearboxes, shafts, and transfer joints of mechanical drive systems. Moreover, the electric drives will distribute the electrical energy generated by the prime power source among onboard sensors, directed energy weapons, and possibly electric guns. Modest improvements in normal drive trains and suspension systems will also occur.

A primary fuel-powered engine (which may be an advanced-concept diesel or gas turbine engine) can be combined with the alternators, electrical energy storage (batteries), and power conditioning units of an electric drive system (Figure 2-11). One advanced concept for primary power is the ultra-high-temperature quasi-stoichiometric, high-pressure-ratio, nonrecuperative simple-cycle gas turbine. Integrating advanced engines with an electric drive that distributes power flexibly to each wheel or track can significantly improve weight distribution and fuel consumption.

Road Building and Bridging

The remote locations of future contingency operations may not be served by modern port facilities. They also may lack extensive road and rail infrastructures. The Army will therefore face the challenges of unloading modern ships without specialized dockside equipment and transporting heavy cargos overland. Although the new drive trains and traction systems described above will assist in transport across rough terrain, heavy logistics service will still require road construction or improvement. Soil stabilization techniques

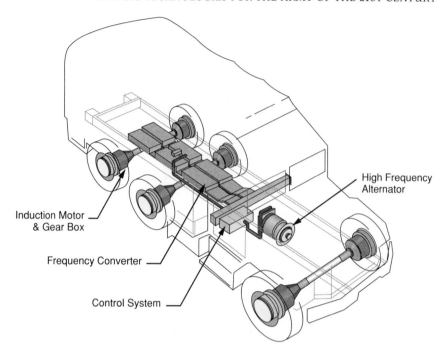

High Frequency
Alternator

Induction Motor
& Gear Box

Frequency Converter

Control System

FIGURE 2-11   Advanced system concept for Army vehicle with advanced engine (not shown), electric drive, and an electric motor at each wheel.

and rapid methods for constructing roads and bridges for heavy loads will require renewed Army attention. Finally, heavy-lift helicopter mobility will continue to be needed for situations where ground vehicles are not adequate. This role is discussed further in the next section.

Rotary Wing Aircraft and UAVs

The roles for helicopters (i.e., manned rotary wing aircraft) in Army operations will continue to evolve, as they have for the past 40 years. But the direction of evolution, as foreseen by the STAR Committee, will differ. The Committee expects new technology to enhance some current roles, such as gunships, forward projection of forces, and supply transport in difficult terrain. Other roles seem likely to wane.

For the mission of forward observation and scouting, the future seems to be running against manned helicopters. Coinciding with the likelihood of increased risk from enemy air defenses is the emergence of a less vulnerable, less costly alternative that does not risk crew lives: C^3I/RISTA UAVs of various sizes and sensor domains. When, or whether, the helicopter is displaced from its observation and scouting role by UAVs depends on a number of factors. For example, it may be possible—although the STAR Committee considers it unlikely—to develop, at reasonable cost, stealth technology that makes helicopters survivable against improved enemy air defense. In any case, the UAV alternative, with its potential for lower cost and less risk of soldiers' lives, should be fully explored.

Other roles for helicopters are likely to increase in the new environment of contingency operations, although there may be competition, or perhaps complementarity, in the long run from UAVs that are custom designed for some of those missions. Two such roles identified by the STAR Committee are logistics transport support over difficult terrain and defense of rear-echelon areas from low-flying aircraft and missile threats.

In operation areas where road infrastructure is inadequate, helicopters will continue, as they do today, to provide logistics transport capability. An engine program to provide the heavy-lift capability that is needed is discussed in Chapter 5; it will be required for either a manned helicopter or UAV solution. Improvements in aerodynamics, robotics, and control technologies will eventually enable the development of an unmanned, tele-operated heavy-lift UAV. The system envisioned by the Airborne Systems Panel would have counterrotating rotors to keep vehicle size and complexity at a minimum, while maximizing the required lift efficiency. This aircraft could carry some 50,000 lb (20,000 kg) over a distance of 200 nautical miles (370 km). Takeoff and landing could be controlled by ground operators at the respective sites; transit would be autonomous. This systems concept is in many respects similar to the Unmanned Air Mobility System now being studied by the Army.

To defend rear-echelon areas against low-flying aircraft or (more likely) cruise-type missiles, an airborne system offers a line-of-sight advantage over ground-based systems. The more compact, lighter-weight sensors, processors, and automatic track-and-recognition systems, coupled with brilliant weapons or directed energy defenses, may enable a helicopter-based air defense station to perform this mission, perhaps as one component integrated with ground-based defenses.

Advanced Armored Fighting Vehicle

Notwithstanding the emphasis that should be placed on indirect-fire systems, the STAR Committee believes that advanced direct-fire armored fighting vehicles will continue to perform the battlefield role currently filled by the main battle tank. There are several reasons for this view. First, the future battlefield will be even more dynamic than today, requiring a system that combines maneuver, protection, and firepower. Hence, the tank seems destined to remain, for some time, the principal mounted system to assault, seize, and occupy defended positions or to counter attacks on friendly positions. Second, the tank's direct-fire main armament, which has a soldier in the loop for target acquisition and fire control, provides highly effective and discrete firepower. Third, a high-velocity gun firing kinetic energy projectiles is of unmatched robustness, especially in the presence of elaborate measures to counter missile guidance systems and chemical (i.e., shaped charge) warheads.

Advances in propulsion, vehicle electronics (vetronics), composite materials, $C^3I/RISTA$, and lethality technologies will substantially improve mobility, command and control, survivability, and firepower. If these technology areas are adequately developed, the armored fighting vehicles of tomorrow can be significantly lighter and smaller yet provide better performance than current main battle tanks. Lethality and survivability will continue to be paramount in future tank design, but its dominant status in ground combat also places a premium on having it present when troops are first sent in harm's way. With reductions in size and weight, perhaps in combination with a modular design that would allow for shipment in readily rejoined sections, it may be possible to transport a limited but critical number of future main tanks with the forces first inserted into a contingency operation.

Among the several keys to reducing the size and weight of future armored vehicles, the foremost is reduction of the armored volume. The crew size can be decreased from four to, at most, two persons; robot loaders, improved sights, automated target acquisition, and stabilized controls will allow the vehicle commander to assume the duties now performed by the gunner. The main gun can essentially be outside the armored hull, further reducing the volume under armor (Figure 2-12). Intensive use of advanced materials can decrease the weight of the vehicle chassis, armor, and engine. Suspension improvements and electric drive can also lessen weight while maintaining performance. All together, a future main armored fighting vehicle incorporating these changes may weigh no more than 60 percent of the current M1A2 Abrams tank. As a result, the strategic deployability of this crucial combat system will improve.

Also, the battlefield mobility of future tanks will be superior to current tanks. The technologies to lighten the overall platform will aid in maintaining or even increasing the power-to-weight ratio.

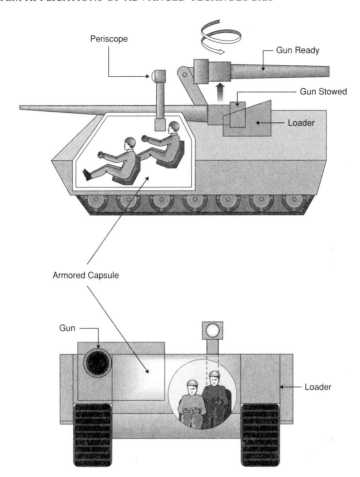

FIGURE 2-12 Concept for an extensible and rotatable gun mount on a direct-fire armored vehicle (battle tank).

Electric drives will be able to provide variable power at each drive sprocket, while offering more flexibility in component placement than in mechanical power trains. Active suspension systems will be able to sense and conform to surface conditions, improving ride quality and permitting increased speeds over rough terrain. Advanced man-machine interfaces for controls and driver assists will also increase the tactical mobility of the future tank relative to the current vehicle.

The operational effectiveness of future tanks will improve through use of command-and-control technologies that link the tank commander with the local C^3I/RISTA assets to identify opposing force elements, moving or stationary, before they can attack, hide, or run. Pattern recognition, expert systems, advanced display techniques, and other information technologies will analyze, interpret, and present this information in a form the commander can use for real-time decision-making on a mobile battlefield. Other technologies that will aid the tank crew include vetronics software; integrated digital mapping, navigation, and position reporting; instrumentation and displays to locate and distinguish friend from foe; and vehicle-mounted multispectral sensors. Commanders of these future tanks will find it easier to be in the intended place on the battlefield, performing their intended tasks.

The main armament of the future tank can remain a high-velocity gun firing a variety of projectiles. As noted, it can be mounted outside the armored hull. Its principal round to defeat heavy armor will probably be a kinetic energy penetrator. Over the next 30 years, the muzzle kinetic energy of the gun is forecast to increase to well over 20 megajoules, or more than double that of current tank guns. This higher muzzle energy can be provided by either electrothermal chemical guns or electromagnetic guns, which will provide muzzle velocities that are 50 to 100 percent greater than velocities achievable with current chemical propellant guns. Because increased muzzle energy will come primarily from increased projectile velocity rather than increased mass, trunnion pull forces will remain tolerable even if the tank weight is reduced. A key enabling technology will be a small-volume pulsed power source. With an electric-powered main armament and electric drive, the future tank may well be an all-electric system.

A guided kinetic energy round will be feasible; whether its increased accuracy relative to the very-high-velocity, unguided round will be worth the cost is unclear at this time.

Survivability of the future tank can be improved despite overall weight reduction. Advanced composite materials and stealth design techniques can make it harder to target by reducing its radar, infrared, acoustic, visual, and dust signatures. The vehicle's smaller size, coupled with a kneeling suspension when the vehicle is at rest, will make it harder to see and to hit. These signature reduction techniques will also make it more difficult for smart munitions to acquire it as a target and guide to it. Conceivably, active defense will enable the future tank to detect and either intercept or divert some munitions used against it, such as relatively slow chemical energy rounds or munitions guided by optoelectronic sensors that can be blinded. Finally, new composite armor can be used that will not produce secondary spall.

These considerations reflect the range of technology applications available for use in future tanks, which can be smaller, lighter, more lethal, and generally higher performing, than the current generation. Because of the combination of capabilities the tank offers, it will not be quickly displaced from its battlefield role. But it can, and should, change markedly to incorporate the newest innovations in those capabilities.

Antiarmor Weapons

The Army will continue to be the service most committed to exploring technologies to overcome enemy armor. Two prongs of this ongoing program must be pursued. The first is to continue the technological advances required to ensure that U.S. heavy forces, when needed in battle, always prevail. The second prong is to give our initially deployed light forces more capability to defend against enemy heavy armor and the other systems that will be brought to oppose them.

It appears that the best of modern armor can defeat any of the currently fielded horizontal-attack antitank guided missiles designed for infantry use. One potential approach to defeating heavy armor is to further increase the size or improve the explosive charge carried by antitank guided missiles (or both). Another approach is to deliver a penetrator having sufficient kinetic energy to defeat the armor. Such a kinetic energy penetrator could be delivered to the target by either a gun or a guided rocket.

A major challenge is to develop a weapon capable of defeating heavy armor but not itself weighing too much to be used by light forces. In practice, rocket systems designed to deliver high-explosive warheads are light but lack robustness against conceivable countermeasures. Rocket systems designed to deliver kinetic energy penetrators are more robust, but they are ineffective at short ranges. (The rocket must burn long enough to achieve the velocity associated with the penetration level of energy.) Gun systems designed to deliver kinetic energy penetrators are quite robust and effective over the range of interest, but they are considerably heavier than rocket systems. The STAR Committee concludes that sustained research on lethal mechanisms will be necessary to ensure that future U.S. light forces can effectively counter heavy armor.

The STAR Lethal Systems Panel envisioned a high-velocity kinetic energy weapon that would be powered by an electric or electrochemical gun. One attraction of this weapon, which should be effective at all ranges of interest to direct fire, is its synergism with the electric drive propulsion systems envisioned for ground vehicles. The kinetic energy gun would fire projectiles weighing 4 to 6 kg accurately to ranges of more than 4 km; the projectile's initial velocity would be about 3 km/s.

A projectile with this energy (about 20 MJ) and range would be effective against the armor threats considered by the Lethal Systems Panel. Active defenses against this type of weapon would be costly even if feasible. A more reasonable defense would be to avoid being targeted by the weapon in the first place; improved Army battlefield C^{3}I/RISTA will make this defense far more difficult as well.

Brilliant Munitions to Attack Ground Targets

The STAR Committee selected brilliant munitions as a high-payoff systems concept for several reasons. Perhaps the most important is that brilliant munitions, whether delivered to near-target range by a "dumb" or "smart" vehicle, will be the key to providing air-deployed forces with sufficient lethality to counter an opposing heavy armored force. Their guidance systems and sensors allow them to be fired indirectly yet still have the accuracy ("zero CEP" or direct-hit capability) to destroy hard targets, including main battle tanks. So they can be used by forces whose own armor is too light to engage heavy forces successfully in close combat.

The Committee also sees these munitions as one warhead option available to various multiple-option weapons platforms. The flexibility to use a brilliant munition with several platforms, each of which can use other munition options for particular purposes, makes both the brilliant munition and its platforms more affordable. Other characteristics of value include effectiveness against a wide variety of targets (e.g., armor, artillery, moving vehicles, command posts, bridges) at distances from short range to deep interdiction, depending on the firing platform and delivery vehicle. Their light weight, small size, and high individual effectiveness will also reduce the logistics burden while allowing the forces using them to stay mobile and outmaneuver an opposing heavy force.

Advanced Indirect-Fire Systems

Technological advances will allow the Army to field an indirect-fire system that is much lighter and more effective than the current 155-mm howitzer or the multiple-launch rocket system (MLRS). Because of its much reduced size, this new system would be much better suited for use by light forces than either the 155-mm howitzer or the MLRS.

A payload weight of approximately 50 lb (20 kg) might consist of brilliant munitions effective against moving armored targets. This two-fold to threefold increase in payload, relative to a 155-mm shell or a submunition of the MLRS rocket, will more than offset anticipated improvements in the armor of the vehicles attacked.

An alternative payload would be used to attack soft, stationary targets. Expected technological advances in rocket guidance should significantly improve its circular error probability. Guidance of the rocket could be based on signals from global positioning system satellites, low-cost inertial measurement units based on micromechanical devices, or a combination of these two techniques. This guidance approach also accommodates a glide-descending and maneuverable trajectory, which increases the maximum range far beyond that achievable by a ballistic trajectory. The gains in range and maneuverability depend on the lift-to-drag ratio of the airframe.

In addition to a lightweight system suitable for use by light forces, another direction for advance in indirect-fire technology is in long-range heavy artillery. One potential systems concept would combine hypervelocity propulsion, to achieve range, with onboard terminal guidance for accuracy. Although hypervelocity projectiles are often discussed for direct-fire antiarmor applications (as in the preceding section), the first fielded systems to use high-velocity electric propulsion (whether electrothermal or electromagnetic) could well be long-range artillery (Figure 2-13). If the range of existing artillery can effectively be doubled, with accuracy maintained or even increased through terminal guidance, the firepower resulting from this technology would be of immense military significance.

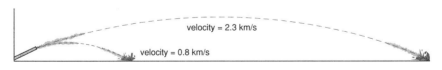

velocity = 2.3 km/s

velocity = 0.8 km/s

FIRE SUPPORT						
		diameter	mass	velocity	range	kinetic energy
Conventional Gun System	(Army)	155 mm	50 kg	0.8 km/s	30 km	15 MJ
	(Navy)	406 mm	800 kg	0.8 km/s	40 km	300 MJ
Electric Energy Gun System		200 mm	100 kg	2.3 km/s	500 km	260 MJ

FIGURE 2-13 Electric gun technology may produce revolutionary advances in artillery range over conventional chemical propellants.

A key benefit of electric propulsion relative to chemical propellants lies not in achieving the most velocity per dollar of energy source but in maximizing the efficiency of the gunbarrel's mass and length relative to the muzzle velocity of the projectile. Both electromagnetic and electrothermal chemical propulsion sustain conversion of source energy into projectile acceleration along the entire barrel length. Conventional chemical propulsion relies on the expansion of gases from the initial propellant detonation. For the same muzzle velocity, a gun propelled by a conventional chemical charge must have a much heavier breech and mountings to withstand the initial explosion. Neither form of electric propulsion is without technical difficulties yet to be fully overcome, but this fundamental advantage argues for continuing current research efforts in both technologies.

Directed Energy Weapons

Directed energy weapons—which use lasers, high-power microwaves, or charged-particle beams as their lethal force—could impose major changes on the character of combat. The STAR Committee suggests that development of such weapons concentrates on antisensor weapons for the purpose of suppressing or damaging visible, infrared, and microwave sensors.

Over the next 30 years, the combat use of laser antisensor weapons is almost certain. Combat use of microwave weapons is probable, but the use of charged-particle beam weapons within this period is unlikely. Heavy-duty directed energy weapons for vehicle kill against aircraft, missiles, and spacecraft are likely to develop first, if at all, as strategic defense systems. They are unlikely to achieve feasibility as tactical weapons of use to the Army within 30 years. In the longer term, as the measure-countermeasure contest unfolds and as tactical responses evolve, the effectiveness of directed energy weapons cannot be forecast with confidence.

Antisensor lasers offer the potential for rapid counterforce defense against electro-optically guided smart weapons and low-altitude aircraft. Within the next decade, lasers with a weight and volume practical for mounting on ground vehicles are expected to reach the output power levels needed for these antisensor applications. Providing the laser steering necessary for targeting a flying threat will continue to be a major challenge. Targeting will become more difficult as target observables decrease while velocities and maneuverability increase. Nonetheless, anticipated improvements in sensing and electronic-processing technologies are likely to match these advances in countermeasures.

Mine and Countermine Operations

For the next generation of land mines and countermine techniques, recent and near-term advances in electronics and sensors are likely to favor the mine side of this measure-countermeasure equation. For different reasons, the Army should vigorously pursue both new mines and new means of countering an opponent's mines. First, in contingency operations a new generation of smart mines—and even of mobile mines in the form of UGVs—can help defend the first-to-arrive U.S. forces, who may be (temporarily) outnumbered and out-weighed (in firepower and armor). On the other side, mines are likely to be used by an opponent to slow a U.S. advance and to inflict casualties in the hope of turning public opinion. Even the threat of hostile mine fields can be a useful tactic, unless U.S. countermine technology is obviously superior to that threat.

Mine Improvements

Given the kinds of threats and contingency operations expected in the future, the STAR panels foresee an increasing importance to the Army of air-deployable "smart mines" and remotely emplaced autonomous mine warheads. These technologies can significantly increase the effectiveness of initially deployed light forces.

New miniaturized sensors and processors, combined with im-proved energetic materials, will enable the development of very small mines that can destroy most ground vehicles. These mines could be so small and numerous that mine clearing would be far more tedious than it is today. Improved electrical sources will allow mines to remain effective for weeks. Simple radio receivers and processors could easily be included to activate or deactivate an entire mine field.

Major advances are likely in the ability of mines to distinguish between friend and foe. The United States and some of its allies al-ready have mines that can distinguish seismic and acoustic signa-tures. In this and other areas, cooperative research with allies will help to leverage limited funding for research in mine and counter-mine technology.

Another major improvement in mines will be their maneuver-ability, whether in air or on the ground. The potential exists for ground-based mines that can fly up and attack low-flying aircraft—perhaps being cued by acoustic signatures from the target aircraft. Low-flying helicopters would be particularly vulnerable to this threat.

The enabling technologies for UGVs will allow mines to become mobile. They can be programmed with sufficient "intelligence" to move into the best position to attack a specific target or to avoid mine-clearing devices. Such mines might be effective against armored vehicles.

UGV technologies may be particularly useful in developing mines that project electromagnetic energy to incapacitate enemy sensors. With supporting intelligence, these mines could be placed to achieve maximum tactical advantage by denying sensor intelligence to an enemy at a specific time (e.g., as a U.S. attack commences). This could be an important new addition to "information warfare."

To realize the full effectiveness of these technological advances, the Army must develop tactics that fully exploit mine capabilities. Simulation before and during battle can be used to develop optimum local tactics as well as compatible geometries and structures for mine fields. Simulation could also be useful in developing concepts for new mines, particularly for mines that have unconventional damage mechanisms or that offer multiple tactical options.

Countermine Operations

Mine detection and clearing have never been efficient or safe tasks. Anticipated mine improvements certainly will exacerbate the problems. Yet new technology offers potential to improve countermine operations as well.

New sensors and sensor data processors, particularly when carried by UGVs and UAVs, offer reasonable hope that most conventional mines can be detected. Both DOD and Department of Energy laboratories are already developing ways to use high-power microwaves to detect mines and to detonate them. Charged-particle beams can be used to do the same. Thermal imaging infrared detectors and laser radar for mine detection are other sensor technologies already in development. Sensor fusion techniques to combine data from multiple sensor domains are being explored. Photon backscatter technology is attractive because it can provide an image of objects whether they are on the surface or buried. In short, advances in sensor technologies and sensor fusion will continue to improve mine detection.

Once the mines in a specified area are detected, they can simply be avoided if the mine density is low enough and the mines are of the same type. For example, technology to silence the magnetic signature of combat vehicles is being explored. The obvious counter to this tactic is to deploy more mines and mines of different types. For example, mines that detect acoustic and magnetic signatures, then home on their target, may be mixed in the same field with simple pressure mines. In this case, either a path through the field must be cleared by detonating or incapacitating the mines (mine-field breaching) or all the mines in an area may be incapacitated (wide-area clearing). For "explosive breaching" of a mine field, air-dispersed powdered explosives are being improved. However, explosive breaching may be ineffective against double-pulse pressure mines and electronically fused mines. Already in development are higher-energy explosives that are safer to store and handle. For wide-area clearing, work is under way to use an expendable UGV decoy that moves in front of advancing forces and mimics the acoustic and magnetic signature of combat vehicles. Chemical and biological agents also may be used in the future to incapacitate mine sensors.

With any of these approaches, computing and artificial intelligence technologies can be used to plan and simulate a mine detection-and-clearing strategy that makes the best use of resources and techniques while minimizing the impact on combat forces. Simulation programs to test tactics, as in cases where mines of different types occur in the same field, are already being used to verify the military usefulness of countermine technology.

Countering Enemy Air Defenses

Many of the Army's potential adversaries can be expected to make substantial progress in their air defenses. For enemy forces with sophisticated weapons and well-trained personnel, this air defense threat will be significant. Some enemy air defense networks will be highly integrated, with embedded target management systems that use artificial intelligence technology. Advanced air surveillance systems will combine (i.e., fuse) sensor data from several locations, quickly detecting all but the most advanced low-observable aircraft. No warning of detection may be possible with advanced LPI (low-probability-of-intercept) sensors.

This sophisticated air defense threat may be susceptible to the type of information warfare concepts discussed previously. Traditional jamming and deception, combined with sophisticated misinformation and other information disruption techniques, might be particularly effective. Yet even less sophisticated threats will have potent air defense systems. Inexpensive, soldier-portable air defense systems will spread to most foreign military organizations as well as to terrorists, cartels, and fanatical religious groups. These groups will threaten not only combat aircraft; they can be expected to attack transport and other unarmed support aircraft operating in rear areas. A particular problem will be that these threat organizations may not abide by conventional rules of engagement and may consider civilian and military medical evacuation aircraft to be fair targets. Information warfare will be less effective against this threat because it will operate from largely autonomous fire units.

These growing threats imply a much more hostile air defense environment for the operation of all aircraft. The STAR Committee views them as indications that the Army will need to move toward a force that relies less on conventional manned helicopters than at present, as discussed in the earlier section on rotary wing aircraft and UAVs.

AIR AND BALLISTIC MISSILE DEFENSE

General Comments

The Army has considerable expertise in developing both air defense and ballistic missile defense systems. For many years, it has played a dominant role in developing ground-to-air missiles for air defense. In addition, from the earliest days of U.S. concern with defense against ballistic missiles, the Army has played a significant role; it continues to do so in its involvement with the Strategic Defense Initiative (SDI).

Several of the STAR panels addressed various aspects of both air defense and ballistic missile defense. Because of the multidisciplinary nature of these topics, a special STAR workshop was convened to integrate the various aspects of the problem and relate them to Army interests and capabilities for the next several decades.

One conclusion of this workshop was that a number of new systems, incorporating new technologies, will be needed. These systems are complex, inevitably expensive, and depend on developments initiated under the SDI. The Army will need to determine which are best developed by others and which are essential for it to develop. The operational need is clearly the Army's, and the Army must participate in defining the requirements and developing the goals, whether the systems are developed by the Strategic Defense Initiative Organization (SDIO), other services, or even U.S. allies.

A key point that must be emphasized is that there will be a number of defense systems operating within a C^3I environment. These systems can, and indeed must, be designed to operate together. This fundamental requirement is an Army responsibility independent of which service or agency develops the systems. Because the Army will operate most (but not all) of these systems, it should be a principal architect of both the systems it will operate and the means to coordinate them all in a larger system of air and missile defense systems.

The Threat Systems

The Army of the future must be prepared to operate in theaters where a wide variety of air and missile systems could be used against it. Achieving a robust defense capability against these threats is both critical and challenging. In particular, the introduction of stealth capability into opposing forces will become a determining factor in fielding an adequate theater air and missile defense.

Potential threat vehicles can fall into any of the following categories:

• *Theater ballistic missiles* (TBMs) have ranges varying from about 100 km to more than 2,000 km. They can fly on elevated, depressed, or minimum energy trajectories (Figure 2-14). They will eventually have some form of penetration aids and pinpoint accuracy.

• *Cruise missiles* and UAVs may be able to operate at altitudes from less than 25 m up to 25 km and at speeds up to several hundred meters per second. They may use stealth technology and electronic countermeasures. Although their operating envelopes are similar to those of manned fixed wing aircraft (see below), they can be much smaller, less expensive, and more numerous than manned aircraft.

• *Standoff tactical air-to-surface missiles* are fired from fixed wing aircraft at ground targets while the launching aircraft remains outside the reach of short-range defenses located near the missiles' targets.

• *Manned fixed wing aircraft* operate at altitudes from less than 100 m up to 25 km and at speeds up to several hundred meters per second. They may use stealth technology, electronic countermeasures, decoys, and infrared countermeasures.

• *Helicopters* operate at comparatively low speeds and at altitudes from ground level up to 3 to 4 km.

Of these threats, the most challenging (as illustrated during the Desert Storm campaign) appears to be the TBM because of its short transit time, high terminal velocity, and small terminal target size. A TBM can carry any type of warhead, from high explosive to CTBW agents, in either unitary or bomblet configurations. In the hands of an aggressor, the TBM is a coercive weapon. The United States, its military services, and its allies will not be credible defenders against aggressive coercion without a defense system capable of countering this threat.

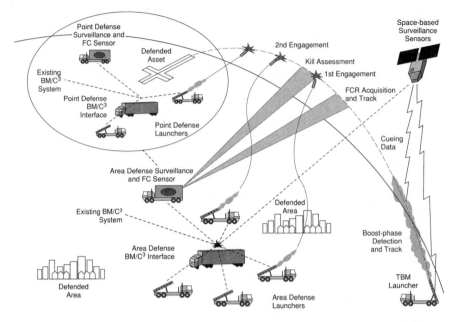

FIGURE 2-14 System concept for two-tier (area and point) theater defense against tactical ballistic missiles. (Courtesy Public Affairs Office, Strategic Defense Initiative Organization.)

In addition to the range of threat systems, an integrated tactical air/missile defense system also has a sequence of action phases: detection, intercept of incoming missile or craft, and counterstrike attack against remaining launchers, airfields, and so on. The larger network must provide threat warning, command and control of interception, and guidance of counterstrikes. Therefore, although the TBM threat may be the most challenging, the larger defense system must be much broader than just a counter to this threat.

Implications for Defense Systems

With such a range of threats to defend against, the rational response is a multiplicity of specific defense systems: a proliferated system for UAVs, an area system for air-breathing cruise weapons or manned aircraft, area coverage for ballistic weapons, and probably point defenses to protect critical installations and respond to stealthy threats that have penetrated other defenses. Any effective solution will involve other services operating with the Army through a joint command. Therefore, the systems used by the various services must be designed to work together, regardless of which service is responsible for developing and fielding the hardware for a particular system.

The implication that emerges is one the STAR Committee wishes to stress: the Army cannot be an effective developer and operator of its share of hardware for this integrated system without participating in the creative analysis of the total problem and the definition of the architecture within which all individual systems must operate. Given the importance of success in this task to future Army operations, the STAR Committee suggests that the Army take the lead in what obviously must be an interservice national effort.

Defense Architecture and Systems

The above line of reasoning shows the importance of a single overall architecture that integrates all of the future air and missile defense systems into a system of systems. The specifics of this integration await definition.

For defense against TBMs, space-based sensors will be used almost certainly to detect missile launch and possibly to track the missiles' trajectories. A framework that combines functions of command, control, and communication (C^3) with battle management must link space-based and ground-based sensors to the system element that controls engagements, commanding the fire units that launch and control the interceptors. A functionally analogous framework will be necessary to defend against air breathers. An early approach to the surface-based elements of such a system could combine concepts successfully applied already in the Army's Patriot system and the Navy's Aegis system.

Many of the systems that will be needed as elements in an integrated "system of systems" for air and missile defense could evolve as enhancements of systems already fielded. The most important requirement is for the Army to work with the other services to arrive at a common plan for the system's architecture. Among the system elements that will be needed are the following:

• an area surveillance, warning, and tracking system to detect and, if not track, at least cue other systems to a TBM launch (a space-based system appears to be the most likely candidate for this mission);
• a similar area system to locate and track hostile air-breathing aircraft and weapons and to assign interceptor systems;
• an effective IFFN system to permit friendly use of contested air space;

• command, control, communication, and battle management capabilities to use interceptor assets for adequate defense of the battlefield or area to be protected; and
• adequate interceptor weapons and local systems for control of interception.

Technologies Applicable to Air and Missile Defense Elements

To achieve an integrated "system of systems," the following advanced technologies would be required:

• High-speed microelectronics are essential to the sensors and high-speed processors.
• Advanced composite materials are needed to construct heat-tolerant, high-speed-flight vehicles that are able to meet the compressed time lines of future intercept systems.
• Bistatic radars may be useful in detecting and tracking stealthy air vehicles.
• Small electronics that can tolerate high acceleration are needed to permit guided projectiles to be gun launched should this form of propulsion prove superior to guided rockets for point defense.
• If guns prove to have advantages over rockets for point defense, pulsed power sources will be needed.
• Multispectral sensors will be essential for extremely fast hit-to-kill interceptors. They may also be the foundation for advanced noncooperative IFFN systems.

SYSTEMS FOR COMBAT SERVICE SUPPORT

Health and Medical Support Systems

Advances in battlefield medicine were discussed above as elements of integrated support for the soldier. The STAR Committee expects that the Army will continue to make major strides in medical and health care capabilities that can be of tremendous benefit to the U.S. civilian medical establishment as well as to the care and treatment of the Army's soldiers. New vaccines, prosthetics, and synthetic tissue replacements, including artificial blood, could be developed for use in Army corps hospitals and disseminated to the wider medical community. Products of biotechnology, such as the diagnostic molecules and enhanced immunocompetence discussed under Integrated Support for the Soldier, will find civilian applications. Even the biotechnology for CTBW defense may find spin-off applications in protecting workers cleaning up hazardous material spills or dump sites.

The greatest area for synergy, however, probably will be in the treatment of trauma patients. Civilian hospitals offer a training ground for trauma specialists as well as for the development of new trauma methods (Figure 2-15). These hospitals have been hard pressed to support trauma centers. Supporting these hospitals with Army personnel can benefit both the Army and the civilian community. Using Army personnel in civilian trauma centers would maintain a trained capability that would be readily available when needed.

In addition, the Army can expect to care for significant numbers of civilian casualties in some future conflicts. The Army should plan to meet this need. This will create requirements for more traditional military medical resources and for other medical skills (such as pediatric specialties). Again, working with U.S. civilian hospitals may provide a synergistic way to develop and maintain the needed capability at reasonable cost, while providing critically needed services to the U.S. population in peacetime.

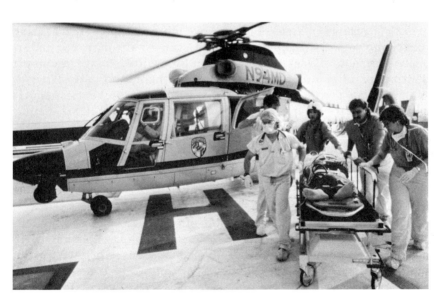

FIGURE 2-15 Civilian trauma treatment centers can foster technology transfer between military and civilian medical professionals during peacetime. (Courtesy of the Maryland Institute for Emergency Medical Services Systems, Baltimore, Maryland.)

Nonmedical Theater Support Systems

Among the systems concepts explored by the Support Systems Panel, the following illustrate how nonmedical combat service support will be affected by changes in technology:

- *Mapping*. Refers to a digital terrain mapping system with a terrain data base system; deployable workstations to update and use the data base; and direct access to terrain data sensors, including space-based sensors.
- *Shelter*. Refers to improved tactical shelters with reduced weight, short erection time, better thermal insulation, some degree of protection from chemical agents, and controlled infrared and radio frequency signatures.
- *Ammunition*. Refers to a computer-based, "paperless" system for control and distribution of ammunition, automated materiel transfer, increased use of intelligent munitions, and higher-energy explosives.
- *Fuel*. Includes an automated fuel tracking system, reliance on a single fuel type, engines designed to run on multiple kinds of fuels, and an armored, low-observable forward resupply vehicle. The Mobility Systems Panel described a vehicle-based hoseline system for delivering fuel to combat vehicles on a dynamic battlefield (Figure 2-16).
- *Maintenance and Repair*. Includes reliability measures such as fault-diagnostic software embedded in the system; an improved failure analysis system; and an efficient system for control, storage, and rapid distribution of modular replacement components and parts.
- C^3 *for Support Systems*. Functions include (1) tracking containers and giving near-real-time locations of stocks in motion, (2) managing supplies and giving near-real-time inventory status and distribution, (3) enabling real-time transportation crisis planning, and (4) controlling spares distribution in theater.

Training Systems

Individual soldiers will be no more effective than the training they receive. Future training and instruction will emphasize the new skills needed by soldiers who will face diverse and unpredictable threats. Future soldiers must understand the capabilities of their equipment and how to use those capabilities in a variety of circumstances. Emerging simulation technologies and individual computer-aided instruction (ICAI) will provide opportunities to enhance soldier performance (Figure 2-17). These tools can be applied to both general training and preparation for specific operations.

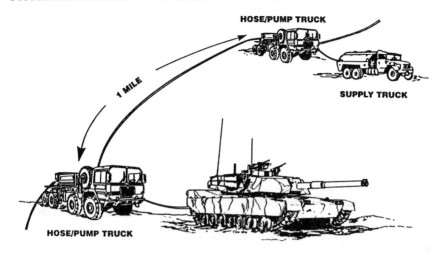

FIGURE 2-16 Concept for a vehicle-based hoseline fuel system. (Courtesy of U.S. Army Tank and Automotive Command.)

Simulation as a Training Technology

Simulation technologies, which are expected to continue improving, will provide an effective means to teach both general system capabilities and their use in diverse situations. The current SIMNET (simulation network) system has been demonstrated to be an effective training device for teaching procedures to small groups of soldiers. Similarly, training simulators such as the Conduct of Fire Trainer (COFT) expand this technology to procedures training for weapons crews and individual soldiers. The Army should continue its emphasis on technologies to improve learning and retention. Areas of emphasis will be geo-specific simulations of combat environments, which will simulate the key characteristics of probable sites of deployment. Advances in computers and data storage during the next decades will vastly improve the reality and effectiveness of simulation training.

ICAI systems will make procedures training more efficient by tailoring instruction to the student's individual needs and progress. As the Army moves to a force structure more dependent on the National Guard and the Reserves, ICAI systems will become more attractive for individual soldier training.

The Army faces the important challenge of better preparing its forces for personal contact with indigenous civilians and for combined operations with allied forces. To be effective in these situations, our soldiers must have a reasonable understanding of the local culture. One way to achieve this understanding is to learn the local language. Future ICAI systems may be able to help Army personnel acquire foreign language skills more quickly.

The STAR Special Technologies Panel forecasts a significant increase in our knowledge of techniques for improving human skills. The Army already has a strong core capability in training technologies; it is appropriately positioned to participate in this field. How-

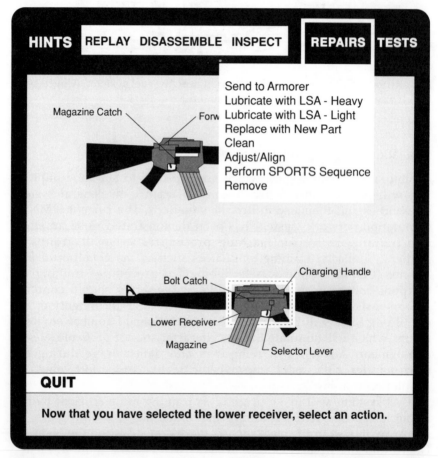

FIGURE 2-17 Computer hardware and software provide the technology base for simulation, modeling, and computer-aided instruction of both individual soldiers and units. (Reprinted by permission, from Mark Miller, Integrating Intelligent Tutoring, in Intelligent Instruction by Computer, Taylor & Francis. Copyright © 1992 by Taylor & Francis. All rights reserved.)

ever, it should bring together its equipment, design, and human factors engineers to work more closely as a team.

Civic Assistance Training

The Army has a special responsibility to support U.S. policies by providing services that do not directly involve combat or combat support. For example, the Army has developed considerable expertise in providing assistance to civic administrators in foreign nations in building their internal capabilities. This assistance has been provided during periods of deployment prior to major combat operations, while combat was in progress (particularly in low-intensity combat operations), and after combat has ceased. The administrative areas in which these services are provided include military forces, civil authority, transportation infrastructure, and medical services.

The STAR Committee anticipates that the Army will need greater capabilities to position personnel in foreign countries in order to establish or reinforce civil authority and critical services. Whether teaching military or medical skills, the training program and the personnel must consider the local culture. The training of Army personnel can be improved to better prepare them for the culture they will encounter.

Technology to help U.S. soldiers acquire foreign language skills will be particularly useful for this purpose. Anticipated improvements in training systems will allow programs to be adapted rapidly to the skills and language of indigenous personnel. Simulation will be extremely beneficial in teaching doctrine and tactics. Two other applicable technologies are the area of artificial intelligence concerned with automated translation of natural language and the computing architectures of neural nets. Within reasonable constraints on vocabulary and context, these technologies may produce a practical means of instantaneous interpretation between languages, thereby overcoming a major impediment to close cooperation between persons who do not speak a common language.

Training of medical and health care personnel will be far more challenging. These same technologies can provide a solid basis for some medical training, but the need will remain for highly skilled Army medical personnel to work with local personnel during and immediately after training. The suggestion made above for assigning Army medical personnel to civilian trauma centers will improve their skills in working with local medical personnel while they maintain their own level of proficiency and stay abreast of changes in technology.

Simulation as a Research, Development, and Analysis Tool

The section above on training has stressed the importance of simulation technology for training, an application area in which the Army has without doubt made substantial use of simulation technology. However, the potential of simulation technology for R&D, which several STAR panels noted (Computer Science, Artificial Intelligence, and Robotics; Mobility Systems; Personnel Systems) has not been explored as fully by the Army. Multiple-unit simulation exercises set in unfamiliar environments or in operational contexts in which the Army has not always succeeded—such as low-intensity conflict or counterinsurgency operations—could contribute to development and test of doctrine and tactics, the effectiveness of prospective weapons and systems, and training of the units involved.

The Mobility Systems Panel noted that SIMNET (the Army's Simulation Network) is used almost entirely for training and not at all for R&D. The panel suggests that SIMNET simulations could be useful input to design decisions on such difficult trade-offs as combat vehicle speed and agility versus armor vulnerability.

The STAR Committee foresees a dramatic increase over the next 20 years in high-realism simulation for large numbers of near-simultaneous interactions of the kind characteristic of the modern battlefield. Furthermore, simulation systems are an area where the United States can expect to maintain a long technology lead. Large-scale simulators that are able to model a modern battlefield with a high degree of similitude can be a technological capability that differentiates U.S. forces from potential opponents. A large-scale simulation capability would allow strategic planners to explore alternatives for U.S. policy implementation, while commanders could use it to explore the means of accomplishing major military objectives, all within the response time required of contingency operations. However, the resources for the simulation (detailed terrain data, data bases of friendly force and opposing force order of battle, logistical support, etc.) must be on call. Those who would use it under emergency conditions must be well acquainted with the system's range of capability beforehand if they expect to rely on it when wargaming is over and warfighting is imminent.

The Personnel System

The expanding diversity of Army missions will increase the need for specialized training and expertise. Finding the time for training will become harder as specialist roles are shifted to reserve units and rapid response becomes critical.

Today, the Army benefits from a buyer's market as its forces are being reduced. Prospective soldiers are typically recruited on the basis of their ability to perform a variety of Army assignments. Psychometric testing is used primarily to screen candidates. The projected demographic trends indicate, however, that in the future the Army will have a smaller pool of individuals from which to recruit. Civilian economic opportunities will continue to compete with Army recruitment and will make retention of Army personnel more difficult.

The STAR Committee suggests that a significant shift in the Army's personnel system can help both recruitment and retention. This changed personnel system would accept a wider range of volunteers but use an increased amount of psychometric testing for classification rather than just selection. Testing that began before individuals joined the Army would continue after they were enlisted. This system would probably require abandoning or curtailing the current practice of guaranteeing assignments prior to enlistment.

The STAR Committee anticipates that remedial treatment of organic physical problems or lack of such cognitive skills as reading or numerical fluency will be available to broaden the pool of candidates acceptable for service. Medical progress may allow correction of diabetes, hypertension, cystic fibrosis, sickle cell anemia, and drug or alcohol dependency. Education and training technologies may allow similar treatment for deficiencies in cognitive skills. Emerging methods in physical training and conditioning may allow enlistment of individuals who today would be unfit for service. This anticipated progress in medical and training technologies can offset some of the demographic trends toward a smaller pool of acceptable candidates.

The STAR Committee also envisions a personnel system that would encourage experienced, trained soldiers to continue in the service. This change will be important primarily because the Army will have a growing need for soldiers who fully understand the broad capabilities of their systems and can use them innovatively, rather than simply apply rote rules for routine use. This expertise can only be developed over time. Further, the historical preference for younger soldiers (under age 30) was based in part on their superior sensory and physical capabilities; these are the capabilities to which advanced technologies can best be applied to augment individual soldier performance.

The envisioned Army personnel system would make continued service by both active and reserve personnel more attractive. It would encourage individual soldiers to remain in one assignment for longer periods, so they could acquire more experience. Research on means of providing feedback to workers on their accomplishments and on areas in need of improvement will improve productivity and motivation. A new area for use of psychometric techniques is in assessment of unit-level skills, interactions, and performance, rather than just testing for individual characteristics. Career counseling for personnel at all levels can make use of advances in psychometric testing and knowledge-based diagnostic analyses to map individuals' aptitudes, acquired skills, and interests into available career opportunities.

3

Technology Assessments and Forecasts

In the request that initiated the STAR study, the first item was to identify the advanced technologies most likely to be important to ground warfare in the twenty-first century. The STAR study included eight technology groups that focused on particular areas of technology. These groups assessed the state of the art and forecast the technology that was likely to be available within 10 to 15 years, so it could be included in Army systems by 2020. All advanced technologies with major Army applications were divided into the following eight technology groups:

- Computer Science, Artificial Intelligence, and Robotics;
- Electronics and Sensors;
- Optics, Photonics, and Directed Energy;
- Biotechnology and Biochemistry;
- Advanced Materials;
- Propulsion and Power;
- Advanced Manufacturing; and
- Environmental and Atmospheric Sciences.

Each group reported its work in a Technology Forecast Assessment (TFA).

After the TFAs for the eight areas had been prepared, a panel drawn from the Science and Technology Subcommittee met to forecast potential *long-term* trends in research that might not produce useful technology until well after the 10 to 15-year time horizon of the other TFAs. The eight area-specific TFAs and the Long-Term Forecast of Research are bound together as a separate volume of the STAR publications.

In this chapter the STAR Committee summarizes what it considers to be the key findings of the Long-Term Forecast Panel and the technology groups, with particular emphasis on their responsiveness to the STAR mandate. The summaries are organized by sections corresponding to the individual reports. For the sake of brevity, much supporting detail has been omitted; the STAR Committee urges readers interested in particular findings to study them in the context of the full report.

LONG-TERM FORECAST OF RESEARCH

Scope of the Long-Term Forecast

The Long-Term Forecast of Research represents the best guesses of a panel of experts on the directions in which technology of interest to the U.S. Army may progress during the next 30 years or more. The principal objective of this report was to highlight significant trends rather than forecast specific technological advances. The forecast panel identified 11 major trends that cut across the traditional boundaries between scientific or technical disciplines. These are discussed below as *major multidisciplinary trends*. In addition, a number of narrower *discipline-specific trends* within specific technology areas will have important consequences for future Army applications. In many cases these trends, which are summarized here, tie in with one or more of the major trends.

Management of Basic Research

The long-term forecast panel agreed that continued support of Army basic research (funding line 6.1) will be necessary if these research trends are to find fruition in Army-specific applications. Budgetary continuity and stability are crucial to achieving long-term objectives.

Major Multidisciplinary Trends

Trend 1: The Information Explosion

The flow of information in preparation for ground warfare and during battle will continue to increase as intelligent sensors, unmanned systems, computer-based communications, and other information-intensive systems proliferate. Data bases and their management software will progress beyond even object-oriented data bases to *third-generation data bases* with new modes of indexing stored data and more intelligence in interacting with the human user of the data base. *Mixed machine-human learning* will team the learning capabilities of a person with the rapid data-processing and analysis capabilities of a computer.

The current limitations to practical application of artificial intelligence may be overcome if an adequate *theory of representation creation* can be developed and *action-based semantics* can be applied to the Army's battlefield information requirements. The information transmission bottleneck on the electronic battlefield calls for data compression techniques; *semantics-based information compression* would address this problem by assessing the value of information relative to the cost of transmitting or storing it.

Trend 2: Computer-Based Simulation and Visualization

Computer simulation of objects and processes, with graphical display of the computer-generated results, gives researchers a potent addition to the more traditional techniques of theory development and experimental evaluation. While computer simulation clearly depends on progress in computer hardware and mathematical algorithms, its growth also depends on understanding the basic principles governing the phenomena to be modeled. Long-term progress in integrating computation with science and engineering may require a broad-spectrum *physical modeling language*, rather than special-purpose simulation environments. Computer studies have already played a major role in modeling the behavior of *nonlinear dynamic systems*. This area of applied mathematics presents both limitations and opportunities for computer modeling of processes important for Army technology. For example, computer modeling will make possible detailed studies of how physical signals, such as light, radar, or sound, propagate in inhomogeneous media, such as the lower atmosphere or through forest canopies. In chemical research, the potential energy surface that characterizes a chemical reaction is a multidimensional mathematical function, which can be modeled and visualized for the researcher. But better methods are needed to approximate the relevant properties of complex molecular systems, and models are needed for reactions of particular interest to the Army, such as combustion or detonation reactions at the surface of an explosive.

Trend 3: Control of Nanoscale Processes

As the features of microelectronic devices shrink to sizes measured in nanometers, new phenomena appear that alter how these devices behave. The particle-wave duality of this quantum world affects both physical and chemical behavior. For example, electron transport, which is essential to all electronic devices, becomes quantized at this scale. Structures no longer behave independently of neighboring structures; quantum mechanical phenomena such as quantum interference, tunneling, and ballistic transport occur. These changes set limits to the miniaturization of conventional semiconductor devices, but they also open opportunities for entirely new devices, such as atom clusters.

Natural biomolecules such as enzymes, or variations bioengineered from them, are likely to provide the first generation of *molecular recognition devices*. Such a device will detect a single molecule of a particular chemical species or with any of a class of molecules with specified structural similarities. Nanoscale chemistry will also control surface reactions, including surface catalysis, through the design and production of layers having an exact placement of component atoms, ions, and molecules.

These new "nanoelectronic" devices will operate at very low voltages and low currents; only a few electrons will suffice to differentiate between the 1 and 0 states of a binary digit. As the technology for quantum-based devices becomes available, subsequent steps will be to integrate them into "molecular" integrated circuits, then into monolithic integrated circuits (wafer-scale integration), which could conceivably have a trillion "devices" on a chip the size of a dime.

Trend 4: Chemical Synthesis by Design

This trend joins with trends 6 and 9 in an even more general trend: in the future, new materials will be *designed at the molecular level* for specific purposes, by designer-engineers using fundamental scientific relations between a structure and its functional capabilities. The realm of engineered chemicals will include both surface catalysts and enzyme-like catalytic molecules, whose specificity depends on their three-dimensional conformation. To support research into these structure-function relations, chemists will need to determine, by experiment and by derivation from quantum chemical theory, the three-dimensional structure of complex molecules, including biomolecules. These structure determinations must be both rapid (on the order of hours or days) and at high resolution (on the order of angstroms).

Trend 5: Design Technology for Complex Heterogeneous Systems

If a system has many components and subsystems that vary markedly in physical and operational characteristics but must act as a functionally coherent whole, it can be considered a complex heterogeneous system. Modern combat vehicles, unmanned air vehicles carrying multiple smart sensors, and a theater air/missile defense system are all examples. At present, the design of such systems is largely a process of muddling through to an adequate result rather than a rational procedure derived from a testable theory. The mathematics of optimization theory can be improved but probably needs to be supplemented, or even supplanted, by other approaches. New approaches are needed for designing systems with *robustness with respect to variation* while taking into account the *costs and benefits of marginal design information.*

Statistical approaches that *seek "least-sensitive" solutions* for a complex design problem hold some promise. But, they currently lack a clear theoretical foundation and may not apply if the system's behavior is nonlinear over its operating range. A radical departure would be to *model the design process* itself, rather than attempting to model the system to be designed. Another area worth exploring is the use of *nonlinear modes of control* for systems whose functional dynamic range includes areas of nonlinear response.

Trend 6: Materials Design Through Computational Physics and Chemistry

This trend combines, within the field of materials science, two other trends: the growth of computer simulation (trend 2) and the design of useful products by application of fundamental relations between structure and function (trend 4). For materials design, these structure-function relations include interatomic forces, phase stability relations, and the reaction kinetics that determine how complex processes evolve. Possibilities of interest to the Army include lightweight (half the density of steel) ductile intermetallics, new energetic materials superior to current explosives and propellants in energy density and safety, materials harder than diamond, and tough polymers with working ranges extending to 500°C.

Trend 7: Use of Hybrid Materials

Also called composite materials, hybrid materials are especially attractive for Army applications because they can be designed for unique and special requirements. For example, the component phases of a hybrid can be altered, or the formation process can be modified, to improve performance in two or more dissimilar functions. The area of greatest technical novelty is that of *smart structures*. A network of sensors embedded in the structural phase of the composite acts like the sensory nerves of an animal's nervous system. A network of actuators allows properties of the structure to be altered, under the control of a microprocessor that reacts to the sensor signals, analogous to an animal brain.

Trend 8: Advanced Manufacturing and Processing

The above trends in designing materials, particularly hybrid materials, will be paralleled by trends in manufacturing *fine-scale materials* (at the scale of individual atoms) and *thin-layer structures*. Chemical synthesis methods such as sol-gel processing will be used, as will methods for controlling process energy precisely, such as laser processing. As *nanoscale devices* (trend 3) become available for sensors and actuators in hybrid materials, smart materials will be synthesized at a molecular level through application of principles such as *self-assembly* and *molecular recognition*. These principles were first studied in biological systems.

Trend 9: Exploiting Relations Between Biomolecular Structure and Function

The principles that relate the functions of biomolecules and tissue structural components to their molecular structure are now well enough understood to be used in designing materials. Among the *potential applications* are new battle gear for the soldier made from lighter and stronger fabrics, broad-spectrum vaccines and prophylactic medicines, sensors and diagnostic devices based on molecular recognition properties, and miniature motors and power supplies based on biological energy transduction mechanisms.

Trend 10: Applying Principles of Biological Information Processing

Biological systems receive, store, duplicate, respond to, and transmit information. The knowledge we have gained about the mechanisms through which this information processing occurs will find practical applications. In the *design of information systems,* capabilities such as pattern recognition and selective abstraction of relevant data may use principles discovered from biological systems. Biological structures, natural or bioengineered, may be *biocoupled* with electromechanical and optoelectronic components (Figure 3-1). At even higher levels of information processing, a growing understanding of the *biological basis for learning and memory* may provide new models and techniques to improve training and performance for information-intensive tasks.

Trend 11: Environmental Protection

The Army will be affected by the general societal trend toward greater concern over environmental effects of toxic materials or disruptions of ecological balances. In the future, the Army will have increased responsibilities for ameliorating past environmental damage and minimizing new environmental contamination or degradation from its operations. Assessing the full impact of hazardous wastes, for example, will require development and verification of accurate models for the transport and fate of the target compounds in soil, air, water, and biota. Better methods to monitor and treat waste materials will be required.

Discipline-Specific Trends

In *electronics, optics, and photonics,* the directions for *advanced sensor technology* include conformal sensors and multispectral sensors, with onboard processors for data fusion and for mission-specific processing such as automatic target recognition. Future Army systems will use an integrated mixture of electronic, photonic, and acoustic devices to process both analog and digital output from a range of sensors gathering electromagnetic, acoustic, and magnetic signals (Figure 3-2). *Active cancellation* techniques will be used to reduce interfering background "noise" and unmask sources of interest. Extensive *communications networking* will require communication links with very wide bandwidths. Allied with the major trend in fine-structure manufacturing (see trend 8) will be advances in *micropackaging* and *minifabrication* of components, subassemblies, and entire nanoelectronic systems (trend 3). Methods for control of optical phenomena will provide faster, smaller, and more powerful architectures for digital data processing as optoelectronic technology expands.

In *aeromechanics,* computer simulations on new supercomputer

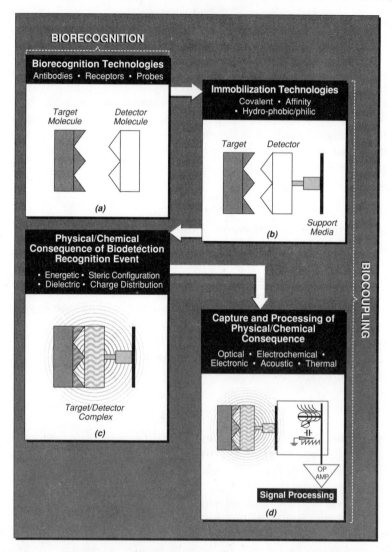

FIGURE 3-1 Events in biodetection: biorecognition and biocoupling. (a) The biologically derived "detector" molecule is capable of a highly specific "recognition" interaction with a target molecule. (b) In the device's configuration, detector molecules are typically immobilized so recognition events can be monitored. (c) When a detector molecule combines with a target molecule, a unique physical/chemical change occurs in the detector-target complex. (d) This recognition-specific change is measured by an appropriate technique, whose output is fed to the signal-amplification portion of the device. Biocoupling comprises the measurement of the physical/chemical change and the subsequent signal amplification.

architectures will allow modeling of rotorcraft vehicles in their operating environment. This greater computing power, combined with advances in computational fluid dynamics, composite structural dynamics, and aeroelasticity will contribute to the goal of complete *aerostructural simulation* (another example of trend 2). Propulsion and control technologies will make *hypervelocity projectiles and missiles* possible. More knowledge will be needed of phenomena associated with hypersonic passage through the lower atmosphere, electromagnetic radiative characteristics of hypersonic vehicles, and the impact and penetration by hypervelocity projectiles against anticipated targets. If unmanned air vehicles become important means for transporting sensors and as brilliant weapons, the Army will require theoretical and experimental data on *aerodynamics at low Reynolds numbers*.

In *molecular genetics*, information deciphered from both human and nonhuman genes will have major implications of interest to the Army. The genetic blueprint information from nonhuman cells will be used in bioproduction of artificial products that mimic natural

FIGURE 3-2 This microcircuit for an infrared detector that requires no special cooling makes possible night-vision equipment for infantry. Future infrared focal plane arrays will combine even more sophisticated image processing in a miniature sensor device. (Courtesy Texas Instruments Incorporated. Copyright © 1991 Texas Instruments Incorporated.)

materials and in the design and production of organisms with new or modified properties. Information about the human genome will yield new methods for preventing and treating diseases or the effects of CTBW agents. Artificial blood, skin, and bone, and perhaps even complex organs such as the liver or kidneys may be replaced by culturing an individual's own cells.

In *clinical medicine*, new instruments and sensors will be used in diagnostic and therapeutic equipment. The miniaturization of sensors (see major trend 3) and sensor data fusion will allow physicians to measure chemical and physiological events at the cellular and subcellular levels as they happen. Army applications include detection of CTBW agents in the field, monitoring of soldiers' physiological condition, and improved diagnosis and resuscitation of the wounded and sick while they are in transport.

In *atmospheric sciences*, high-resolution *remote sensing* of meteorological conditions will provide the data to initialize and validate *computer models* of the atmosphere on small spatial and temporal scales, for which the Army has special need. The validated computer model can then be used to improve sensor placement. By repeating this cycle, the sensor data-gathering and computer modeling activities will complement one another. The result should be increased understanding of small-scale weather conditions, including fog and cloud physics and more accurate representations of turbulence.

In *terrain sciences*, sensor technology and information processing are again important, for both *automated extraction of information from multiple imaging* and *three-dimensional representation of terrain data*. A key addition to existing terrain data capabilities will be a near-real-time system to analyze and map changes in terrain surface conditions and trafficability. Such a system would use sensor data on rainfall, soil moisture monitors, and computer modeling of soil properties based on hydrologic and atmospheric conditions.

COMPUTER SCIENCE, ARTIFICIAL INTELLIGENCE, AND ROBOTICS

TFA Scope

The Computer Science, Artificial Intelligence, and Robotics Technology Group assessed the following technologies:

• *Integrated system development* includes system development environments, design languages and compilers, problem-solving strategies, simulation and optimization (in development), and the mathematics for representing and managing variation.

• *Knowledge representation and languages* includes mathematical representations of information and special-purpose languages, such as battle control languages.

• *Network management* concerns the management of multiple processors that pass digital data or other information (such as voice messages) to one another through interfaces.

• *Distributed processing* is the execution of a computation (a program or a number of computationally independent programs) on two or more processors. Usually the processors are part of a network.

• *Human-machine interfaces* include graphic displays, keyboards, control consoles, pointing devices, printers, audio outputs, and other means by which a computer or peripheral communicates to the human user or the user communicates to the machine.

• *Robotics* includes stationary and mobile systems, airborne or ground-based, that are controlled by onboard computer programs. They may be (1) autonomous, (2) supervised by an operator but operating autonomously for routine operations, or (3) under continual operator control (tele-operated). Their mission may require sensors and communication capabilities only, or they may have advanced processing and even weapons capabilities.

• *Technologies to monitor* are areas in which the Technology Group thought that nonmilitary R&D would lead the way and the Army could profitably use the results without funding research itself. These areas include machine learning and neural nets, data base management systems, ultra-high-performance serial and parallel computing, planning, manipulator design and control, knowledge-based systems, and natural language and speech.

Technology Findings

General Findings

The battlefield of 2020 will use millions of computer systems and components. These systems, ranging from tiny microprocessors embedded in weapons to mobile command-and-control centers, will be ubiquitous, critical, and essential. They will be interlinked by a wide range of communications media.

The effectiveness of individual *soldiers* in the future will be enhanced by computational tools that give them constant access to command-and-control centers, help them navigate, monitor their physical condition, and provide an instant source of up-to-date knowledge in the form of "smart manuals."

Data bases, nearly instantaneous communications and analysis, intelligent decision aids, and multisensory information displays will provide *commanders* with an unprecedented awareness of the battlefield (Figure 3-3). While the potential for a good commander to affect the outcome will be multiplied, so will the potential for command errors to prove disastrous.

For *logisticians*, the most significant changes will be in the planning and control of logistics operations.

For *strategic planning*, warfare fought with computers and unmanned systems may become at once more common and more threatening. The publics of advanced nations may find war more acceptable if the number of casualties can be kept low. The rewards of aggression may be higher because of the aggressor's ability to exploit a temporary advantage in system sophistication.

Computers, data bases, and software will themselves become targets in warfare. Our computational resources must be protected while exploiting any vulnerability in the opponent's systems. This *computer*

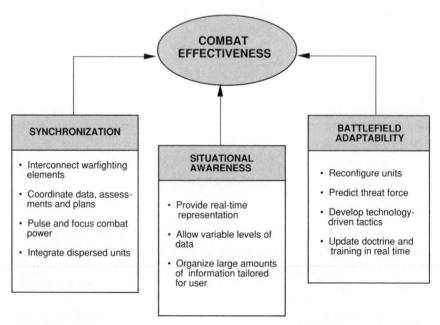

FIGURE 3-3   Technology based on computer science will help commanders improve combat effectiveness.

warfare will involve at least four components: information security; injection of, and protection against, electronic viruses; sabotage; and exploiting the computational predictability of the opponent's systems.

Integrated System Development

The Army's development problems dwarf those of any other U.S. organization, governmental or private sector. For example, the Army will need to learn how to structure the simultaneous development of (1) the systems themselves, (2) the civilian surge capacity to produce them in large numbers without major peacetime investment, and (3) the doctrine to use their often revolutionary capabilities.

If the individual technical areas that contribute to integrated system development are considered separately (e.g., software engineering, electronic systems, or mechanical design), the Army could follow the lead of private sector developments. However, where these technologies relate to one another and where they affect tactics and training, the Army will need to lead.

Civilian developments can accelerate the introduction of technology into Army applications and reduce costs. Opinions vary on how much of the computer and systems hardware produced for civilian markets is too fragile for direct Army use. However, some civilian items will not meet minimum functional specifications, so it will be necessary to analyze carefully the effective use of components and systems developed for the commercial market. In some cases it may be necessary to redevelop them to military specifications; in many others (e.g., microcomputers), a preferable route is to provide an environment in which commercial items work well enough.

The 2-year cycle for computer obsolescence and the vulnerability of unmanned systems to countermeasures will make it infeasible to maintain constant fielded superiority over every *potential* threat. Systems to meet potential threats will need to be designed but left unfielded unless the corresponding threat materializes.[1]

The successful use of *high-level design languages*, including compilers to translate high-level design into detailed descriptions, will depend on satisfactory answers to two questions: Does the description

---

[1]The STAR Committee adds that simulation training would become essential under this scenario, so that unfielded systems could be introduced with minimal delay.

accurately reflect what the designer wanted it to? Is the high-level description correctly translated into an acceptable implementation?

The Technology Group forecasts that software design will remain a difficult problem in 2020, but very-high-level languages will have shifted most software development out of the hands of programmers and into those of subject-area experts (Figure 3-4). The Group also forecasts that by 2020 the Army will have the infrastructure, design tools, languages, and silicon foundry engineering to deploy to forward maintenance depots a silicon compiler for automated design and production of VLSI chips.

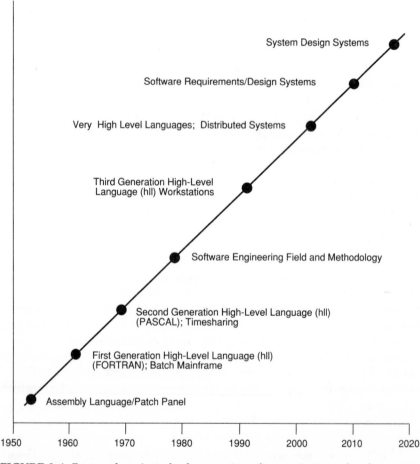

FIGURE 3-4 Past and projected advances in software systems development.

Problem-solving strategies in artificial intelligence are beginning to progress beyond heuristic classification, which seeks a valid solution from a fixed set of possibilities, to heuristic construction, which creates a complex solution from incremental subsolutions. To apply this emerging technology to the Army's complex design problems will require further advances in (1) conceptual modeling of application domains, (2) abstraction hierarchies, (3) representations for partial design information that capture dependencies, and (4) problem-solving strategies for high and low levels of abstraction.

In the area of *design simulation* and visualization tools, by 2020 high-resolution simulations will be calculable from first-principle nonlinear equations and physical relationships. Real-time, interactive simulations of complex systems and operational environments will be achievable by networking thousands of individual simulators.

Optimization programs will be available to vary the parameters of a simulated design to seek the best set of design parameters. However, this optimization approach will still depend on having an initial design to simulate; the form in which a design problem is represented can control its solutions.

System developers must reason about sets of objects under sets of conditions. Mathematical representations of variation, both probabilistic and nonprobabilistic, will therefore be needed for integrated system development (as well as for other application areas). The Army will need to participate in this research because the battlefield, as a source of *deliberately antagonistic variation*, is unlike the environment faced by private sector developers.

Knowledge Representation and Languages

Advances in knowledge representation will increase the reliability of software by providing formal structures and mathematics to describe key information about the battlefield, such as terrain, the degree of certainty about the enemy's forces and intentions, and sets of potential outcomes.

The search for new mathematical representations of knowledge is being driven by the computer. Often, the needed representations do not exist yet.

The Technology Group envisions a *battle control language* that will give commanders control of computational power analogous to the control that current spreadsheet packages give users to "program" their own calculations and tables. The high-level language will use statements that look like operations, orders, unit TOEs (Tables of Organization and Equipment), and map graphics. The language will let commanders control, interrogate, and understand a nearly instantaneous information flow about unit status and logistics. Incoming intelligence will be correlated and displayed in seconds. Continuous simulations, which will run in the background, will be used to test alternatives. Broad mission orders from the commander will automatically generate implementing instructions to units.

Network Management

The networks of 2020 will carry voice and data at high data rates. Connection into the network will be available anywhere in the field. However, some of the unsolved problems of network management today are likely to continue. These problems include delay, redundancy, and priorities.

Managing communications bottlenecks within networks and across networks connected by gateways will be a critical task. Problems arise when different networks with different access schemes, protocols, and security levels communicate with one another or pass message traffic on to other networks.

The Army has a useful role to play in solving these network management problems. It can offer prototype and experimental environments where new approaches can be stress tested.

Distributed Processing

Today, the level of distributed processing has advanced to the point that a new application process needs to interact only with the operating system or network protocol, one step above interaction at the hardware level. The Technology Group forecasts that this level of required interaction will advance by 2020 so that applications will interact at a level of abstraction (meaning) that is far above the hardware level. This level of interaction will enable more powerful application programs for sensor fusion, situation assessment, operations planning, and sophisticated modeling.

The technology to support this future level of distributed processing will include massively distributed worldwide networks; dynamic, real-time protocols for distributed operating systems; and distributed object data base management systems.

The Army's use of distributed processing will face more difficult obstacles than occur in most civilian environments: (1) continually varying prioritization of processes; (2) robustness when large parts of the distributed system disappear without warning; (3) a vast range of data types, inference types, and hardware; and (4) accommodation of several levels of security and access.

Human-Machine Interfaces

Given the potential for information overload, a crucial requirement will be to provide the human users of a system, particularly commanders, with the information they need without overwhelming them. The difference often depends on good interface technology.

In the area of *human factors*, technological advances will occur in visual display techniques, force-feedback controls, information presentation optimized for low data transmission rate, and workload optimization for control of multiple systems by a single operator.

Heads-up displays, which project an image from a lipstick-size tube onto eyeglasses, are just appearing as commercial products. By 2020 stereo heads-up displays will be standard.

The Technology Group projects that by 2020 interface media and modes will be customized to the user's job, expertise, and personal preferences. One operator will be able to control multiple systems simultaneously and efficiently. Controls will switch between pure program control, operator monitoring, and operator control, according to circumstances. Structured voice will be the standard mode of entering responses to option menus. Other input modes may include analysis of facial and body gestures from TV images or direct monitoring of physiological responses (although the semantic content of such input will be far less sophisticated than linguistic expression).

In the *hypermedia* area, users will be able to navigate consistently through many different kinds of information, including drawings, photographs, video, synthesized voice, and diverse textual formats. The technology and human factors expertise for hypermedia is just beginning to emerge.

The Technology Group forecasts that users of map information in 2020 will be able to switch freely between, or superimpose, symbolic data and simulated or real scenes. Synthesized speech will supplement visual displays. Computer-aided drawing will convey nonverbal information.

Robotics

The core weapon of land war in the twentieth century has been the tank. The core weapon in the twenty-first century may well be the unmanned system, operating mostly under computer control with human supervision. Robot systems may be classified according to the level of continuous operator control during the system's operation. *Fully autonomous robots* perform their tasks with no human interaction following mission assignment (i.e., after they are programmed for a task). *Supervised robots* can perform most of their assigned task autonomously but require interaction with their human operator from time to time or when special situations arise beyond their programmed capability. *Operator-controlled robots* require interaction with a human controller at frequent intervals. While one operator can control several (perhaps many) supervised robots, an operator-controlled robot in practice requires the full attention of its human controller. The term "tele-operated" may apply to either supervised or operator-controlled robots, as distinguished here.

The Army's requirements for successful battlefield robotics are unparalleled in the private commercial sector. Robots are feasible on production lines because variability in the environment and range of stimuli can be contained. However, variability on the battlefield is uncontainable to the extent that the enemy can affect it. Even by 2020, unmanned systems will probably still be less capable than manned systems. They will be useful because, if properly designed, they can be far more numerous than manned systems. The highest payoffs from battlefield robots will come from putting large numbers of sensors in places where soldiers should not go and from integrating the information from the sensors into a coherent picture.

The Technology Group forecasts that a more relevant model for conceiving of battlefield robots is the land mine rather than the human soldier (Figure 3-5). That is, military robots will evolve as "smart mines," with increasingly sophisticated sensors, weapons, and modes of propulsion, rather than as "mechanical foot soldiers." The sensors, weapons, and propulsion methods used by these robot mines will differ greatly from those used by soldiers.

To defeat enemy attempts at deception, battlefield robots will have to integrate a wide variety of sensor information. The Army of 2020 will have vast requirements for signal processing from a single sensor, sensor fusion, and sensor integration.

For robot weapons, the most practical concept may be explosively propelled projectiles that achieve armor-piercing velocity with low weight and cost. By 2020 single-missile robot tubes, hidden in ground cover, are likely to be more secure than missile batteries mounted on vehicles.

The trade-offs among range, cost, and flight time of robot weapons, whether from airborne or ground-based launchers, imply that a mix of systems is preferable to a more complex, all-purpose system. A typical unmanned system should be specialized. Properly con-

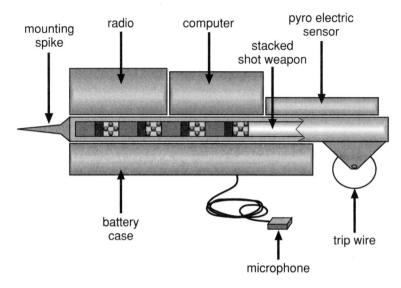

FIGURE 3-5 Concept for a simple military robot that can hold ground.

ceived, battlefield robots can be inexpensive and quickly developed, yet they will remain vulnerable to countermeasures and can rapidly become obsolete.

The mechanical issues of robot vehicle mobility are quite different for air and ground systems. Large robot air vehicles, whether autonomous (e.g. cruise missiles) or tele-operated (drones and remotely piloted vehicles), are well established already. The cost and bulk for terrain mapping and related navigational computation for airborne vehicles will continue falling. By 2020 the Technology Group foresees the possibility of building actuators, sensors, and computers on a single silicon chip.

Mobility for small ground systems over natural terrain is a major development challenge. The simplest mode, mechanically and computationally, is leaping followed by reorientation. The use of mechanical legs for walking or running motion will be more difficult, although the Technology Group expects both running and walking legged robots to be in use by 2020. Other options are wheeled or tracked vehicles.

The packaging of huge numbers of components into a mission-specific configuration is another key technological requirement for battlefield robotics to succeed. The Army may need to play an active role in supporting or conducting research in packaging.

Technologies to Monitor

Machine learning.   Recent research in this area has produced a large number of software systems, most of which are tailored to one learning paradigm. During the next 30 years, integrated learning systems, designed for general learning problems, will gain more research attention. These systems will be able to adopt different learning strategies, depending on the problem at hand. However, there is still no general theory of learning. If the strategy is poorly suited to the problem, unwanted or incorrect generalizations can result. Another difficulty is that learning systems may be difficult to debug or to modify for changed circumstances.

Neural nets.   The Technology Group views neural nets as a particular mechanism for machine learning. Their architecture seems closer to biological computation (animal and human brains) than do conventional programmed architectures. But the Group sees no convincing argument to conclude that this resemblance will make neural nets superior in "intelligence." They have, however, achieved some spectacular successes in pattern recognition, which could make them useful to the Army.

Data base management systems (DBMSs).   Existing relational DBMSs will be superseded by knowledge-based, hyperdocument DBMSs, which will be able to represent complex data structures such as tables, large text documents, images, and maps. Also important will be a move to fully distributed DBMSs with automatic updating, maintenance, and dynamic optimization of storage location.

Ultra-high-performance parallel and serial computing.   The modeling and simulation needs of the Army will continue to make use of the latest and fastest computers. The two major applications are scientific simulation as part of the development process and simulation of combat activities or wargames. Parallel architectures seem to be the only promising route for continuing the past rate of growth in computing power. Whether the conventional supercomputer design, which is based on vectorization of the computational problem, or radical alternatives such as massively parallel architectures is optimal appears at present to depend on a case-by-case fit of the problem to be solved, the software program that solves it, and the machine architecture.

Planning.   By 2020 planning technology will be capable of producing complex plans in complex domains. The most dramatic progress will be in the breadth of knowledge that is brought to bear in generating or revising a plan. Memory of past events, including plan successes and failures, will be used. Contingency alternatives will be explored to greater depth before an optimal course of action is selected. Revision and adaptation will be faster.

Manipulators. By 2020 manipulators will have more than 10 degrees of freedom, with capacity-to-weight ratios 10 times the current state of the art. They will be fully modular and mission-configurable.

Knowledge-based systems. By 2020 knowledge-based systems will begin to approach a general problem-solving capability, although they will still be restricted to one class of problems. They will be able to handle broader, more general knowledge representations; model time more richly; perform inferences under real-time constraints; and perform problem solving that is distributed across multiple processes, processors, and physical sites.

Natural language and speech. Although natural language processing will be able to understand report-length texts or long messages in well-defined domains, reliability will remain the essential issue. Natural language technology is most likely to be applied where error detection (by human review of the machine product) is possible. It may find use in "pretranslation" systems or in watchdog systems that scan large volumes of material for items to be brought to an operator's attention.

ELECTRONICS AND SENSORS

TFA Scope

The Technology Group on Electronics and Sensors assessed the following areas of technology:

- *Electronic devices* include advances in monolithic microwave integrated circuits, superconductive electronics, vacuum micro devices, computer memories, application-specific integrated circuits, analog-to-digital converters, digital signal processing microcomputer chips, and wafer-scale technology.
- *Data processors* include electronic subsystems that act as signal processors and target recognizers. The emerging technologies here are multiprocessor computing and neural networks.
- *Communication systems* include communications satellites and

other platforms for communications nodes. They also include technologies to provide communications security and robustness.

• *Sensor systems* include UAV detection radar for surveillance of moving ground targets, airborne detection and recognition radars for stationary targets, acoustic array sensors, magnetic sensors, air defense radars, and space-based surveillance and target recognition radars.

Technology Findings

The findings reported in the Electronics and Sensors TFA include general findings, those specific to the technology areas specified above, and summary findings on three high-impact electronic technologies.

General Findings

Land warfare is likely to evolve, as air and naval combat have already, toward long-range weapons, increased depth of combat, and increased reliance on stealth, electronic countermeasures, and mobility. The key to this evolution is the development of electronic sensing and target recognition systems that can operate beyond the visual horizon.

Electronic Devices

Silicon will remain the *bulk semiconductor material* of choice for most applications for the foreseeable future, if only because of the high industry concentration of development resources and manufacturing capability committed to it. Although some significant evolutionary improvements in silicon semiconductors will continue, the next decade will begin to see the performance of silicon devices limited by intrinsic properties of the material.

In special areas, greater performance gains will occur because other materials are far superior to silicon. For example, gallium arsenide (GaAs) is becoming the material of choice for high-frequency transistors. Silicon carbide or diamond may emerge as the semiconductor materials of choice for devices operating at high temperature and high power.

In the area of *thin-layer semiconductors*, the well-established technology for chemical vapor deposition of thin silicon layers on silicon substrates is being augmented by new techniques for growing single-crystal layers from the vapor phase. These new techniques allow deposition of highly uniform layers of solid solutions (such as

GaAlAs) as well as elements and binary compounds. It is now possible to fabricate extremely complex multilayer structures whose properties differ dramatically from any bulk material. The long-term impact of these techniques will be significant improvement in device performance.

The recent discovery of materials that are superconducting above 77 K has renewed interest in *superconducting thin films* for high-frequency analog and digital circuits. Films with sharp superconducting transitions, high critical-current densities, and low microwave losses have been obtained in the laboratory. Superconducting thin films enable major performance advances in a variety of microwave, radio frequency, and digital logic applications.

Monolithic microwave integrated circuits (MMICs) have been made possible by the developments described above in high-quality semiconductor materials, new thin-film deposition techniques, and improved lithography. MMICs provide small-signal amplifier and power amplifier components for applications in the range from 10 to 100 GHz (Figure 3-6). They will enable phased-array radars, signal intercept systems, and communications terminals to be built with far

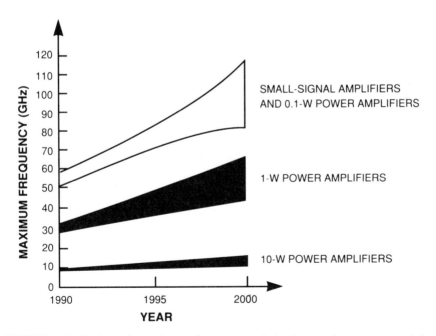

FIGURE 3-6 Projected maximum frequency of single-transistor commercial amplifiers.

smaller size, weight, and cost compared with conventional hybrid technology. This new technology will be crucial in developing the following systems: UAVs with multi-use apertures for both ground surveillance radar and electronic intelligence (ELINT) receivers; space-based imaging radars; covert, beyond-line-of-sight communications "manpack" terminals; and extremely-high-frequency (EHF) terminals for air-to-air and air-to-satellite links.

Micron-size vacuum transistors have become possible with the development of reliable cold cathodes with high current densities. Vacuum transistors, which operate on the same principles as traditional vacuum tubes, will have high-frequency and high-power capabilities beyond those of semiconductor transistors. They can be used to develop radar and communications systems at frequencies and power levels not attainable with current solid state technology. They also could replace the traveling wave tubes now used, with substantial reductions in the size, weight, and power consumption of the system. A principal advantage of these devices is their robustness to damage from electromagnetic pulse (EMP) associated with nuclear blasts or video-pulse directed energy beams.

The advances described above in semiconductor materials, thin film techniques, and circuit miniaturization will all contribute to the emergence of *advanced electronic devices*, which are still in the concept stage. For example, these evolutionary advances in microelectronics will aid in developing the optoelectronic circuitry needed for revolutionary advances in optical computing and neural networks (see Optics, Photonics, and Directed Energy). Similarly, they will be needed in the field of bioelectronics, including biosensor coupling and, ultimately, biocomputing systems (see Biotechnology and Biochemistry).

Computer memory chips will continue to increase in capacity; in silicon technology, both direct random access memories (DRAMs) and static random access memories (SRAMs) are projected to gain nearly two orders of magnitude in bits per chip during the period 1989-2000 (Figure 3-7). Cost will also continue to decrease. In addition, high-speed memories (2 nanoseconds compared with 10 nanoseconds for CMOS SRAMs) based on GaAs or silicon carbide, which are now in development, should show a corresponding rise in capacity.

An *application-specific integrated circuit* (ASIC) is a single-chip substitute for a subsystem previously assembled from a number of simpler, standard chips. The use of ASICs in place of subsystem assemblies increases reliability while reducing the number of components and lowering the production cost, weight, and power required. Generally, however, the system development cost increases. The least costly type of ASIC, the programmable logic device, is already well

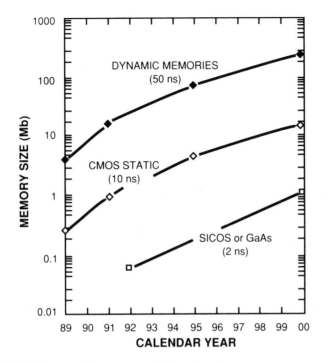

FIGURE 3-7 Forecast for memory chip technology.

established. It can be programmed in the laboratory in a few hours. Gate-array or standard-cell ASICs, which must be programmed at the mask level during circuit manufacture, require 6 weeks to 6 months to produce and cost about $100,000. In the next 10 years the capability limit for gate-array ASICs will grow from about 10,000 gates to around 500,000 gates at a 50-MHz clock speed.

At the upper end in both cost and capability is WST (wafer-scale technology), which can implement an entire system on a single substrate. The cost of a WST chip is about $1 million, and production time is about 1 year. WST offers the advantage of eliminating many of the separate fabrication steps required to implement a digital system. The potential of WST can be indicated by designs achieved or in development now (Figure 3-8). A fast Fourier transform unit on a wafer 7.5 cm in diameter, demonstrated in 1986, had a throughput of 300 million operations per second (MOPS). A WST design on a 12.5-cm wafer, under development in 1990, is expected to achieve 2 billion

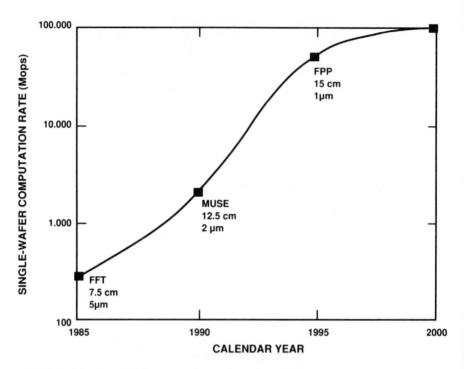

FIGURE 3-8 Capabilities of wafer-scale technology.

operations per second. By the mid-1990s, increased wafer size, re-duced feature size, and faster clock rate will together increase WST computation throughput to 50 to 100 billion operations per second. For potential military applications, more relevant measures are the computations per unit size, weight, or power; WST seems capable of achieving 100 MOPS/cm³, 200 MOPS/g, or 3,000 MOPS/W.

Analog-to-digital converters (ADCs) provide the connection between the analog world of sensors and the digital world of data processing. As sensor technology expands, either in bandwidth (as in radars) or focal plane size (for optoelectronic systems), the requirements also expand for wide-band ADCs with high dynamic range. ADC technol-ogy is being rapidly advanced in frequency and precision capabil-ities by commercial sector interest, particularly for high-definition television. However, military applications also require high dynamic range to accommodate their wide range of signal levels. Research is under way, with the support of the Strategic Defense Initiative Organization (SDIO), to develop monolithic (single-substrate) ADCs suitable for military use, including radiation hardening.

A *digital signal processing microprocessor* (DSP) is an integrated circuit similar to the microprocessor in a high-performance microcomputer but designed to be optimal for signal processing applications such as filtering, spectral analysis, and convolution. A new development in this area is the *DSP microcomputer*, which can sustain an average computation rate close to the peak rate required by a typical DSP task. These single-chip devices can be integrated into compact systems.

The Technology Group expects the DSP microcomputer to attain faster clock rates and smaller feature size over the next five years (Figure 3-9). Designs to exploit parallelism, such as multiple processors on a single substrate, will appear. In ten years GaAs technology will enable 100-MIPS DSP microcomputers. Fiber optics will provide gigabit interprocessor communications. Arrays with up to 1,000 parallel processors will become available, although software design methodologies to support this level of parallelism will also need to be developed. Among the implications of DSP microcomputer chips

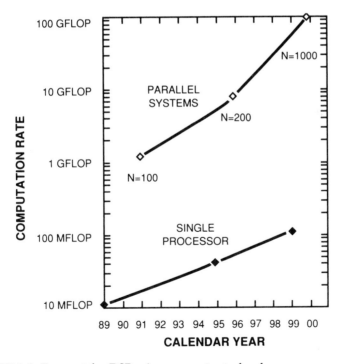

FIGURE 3-9 Forecast for DSP microcomputer technology.

for ground warfare will be increased sophistication, while reducing the size, of systems that rely on signal processing. Examples include compact smart weapons with onboard target recognition, communications systems with advanced low-probability-of-intercept technology, speech recognition in command and control systems, and radar and sonar systems with sufficient sophistication to detect stealthy aircraft and quiet submarines.

Major advances are occurring in *electronic design automation*. The key requirement is to have the data output from one computerized design or fabrication step in the multistep process be directly interpretable by the next step in the process. Two data format standards for this purpose are the VLSIC (very-large-scale integrated circuit) Hardware Description Language (VHDL), developed under DOD auspices, and the Electronic Design Interchange Format (EDIF), which is widely used in the commercial electronics industry. Software programs that synthesize data paths and entire circuits are available, although they provide quick turnaround at some cost in performance and silicon "real estate" efficiency. Other important tools are logic and circuit simulators. Much work is being done on developing integrated sets of tools for electronic design automation. (This need for integrated system design environments is also addressed in the Computer Science, Artificial Intelligence, and Robotics TFA; see section above.)

Data Processors

Increases in computing power in the near term will result from a new generation of VLSICs, multiprocessor computer architectures (parallel-processing supercomputers), increased use of GaAs and other high-speed semiconductor materials, and the technology for reduced instruction set computing (RISC). These changes will increase both numerical computing power (i.e., millions of floating point operations per second) and symbolic computing power (Figures 3-10 and 3-11).

A *neural network* is a computing architecture that performs highly parallel processing with a large number of simple processing elements (called the neurons). The neurons may be sparsely or densely interconnected. The potential advantages of neural networks include high-speed processing through parallelism, robustness to individual element failures, and compact hardware implementation of entire networks as VLSI chips.

Current realizations of neural nets are almost entirely in the form of simulations on a standard digital computer. Hardware realizations are still experimental, although rapidly maturing to the point

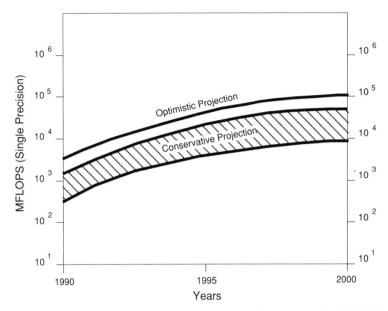

FIGURE 3-10 Projection for numerical computing power (1 MFLOPS = 1 million floating point operations per second).

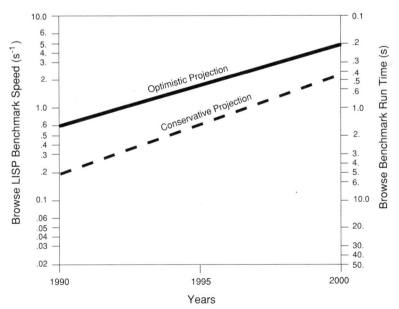

FIGURE 3-11 Projection of symbolic computing power (in terms of time required to perform the "Browse" benchmark program).

where neural net chips will be included in commercial information-processing systems. With sufficient commitment to their hardware implementation, neural nets could bring revolutionary changes to military systems. Their advantages of high-speed processing by rugged, compact hardware with little dependence on software would be significant for such applications as brilliant weapons, autonomous systems (UAVs and UGVs), automatic processing of sensor data, image processing, and adaptive signal processing and control. (Neural network realizations using photonic and optoelectronic hardware were discussed by the Optics, Photonics, and Directed Energy Technology Group; see abstract below.)

Communications Systems

The Military Satellite Communications (MILSATCOM) architecture for the period after the year 2000 calls for *communications satellites* with the current ultra-high frequency (UHF) and super-high frequency (SHF) services, plus a robust/survivable segment and a complementary capability for augmentation and restoration.

The *robust/survivable segment* of this future MILSATCOM will operate in the extremely high frequency (EHF) range of 20 to 44 GHz. To support antijam, antiscintillation, and covert communications, it will use wideband, spread-spectrum techniques and autonomously adaptive uplink antennas. Low, medium, and high data rates must be supported, and agile uplink/downlink beams will be used to serve widely separated users concurrently. Satellite cross-links will provide worldwide connectivity without ground relays. The Technology Group identified the key electronics technologies required to accomplish these operational goals in a payload with substantially lower weight and power requirements than current technology. The new technologies include MMICs, VLSI processors, direct digital frequency synthesizers, and WST ASICs.

The *augmentation/restoration* satellites for the future MILSATCOM will be used to increase or replace critical coverage in a timely manner. Using the new technologies mentioned above, they can have many of the operational features of the robust/survivable segment mentioned above, except for supporting high data transmission rates.

Sensor Systems

The continuing development of smaller, more capable processors will benefit all radar systems but will be particularly significant for *UAV-based radar systems*. For example, UAV radar systems for the detection of moving ground targets can be valuable adjuncts to the large systems, such as JSTARS (Joint Systems Target Acquisition Radar System), carried by manned aircraft. A current example is the DARPA-sponsored AMBER UAV, whose Ku-band radar includes a programmable processor to interpret raw radar data into moving-target reports. Moving targets the size of tanks or larger can be tracked out to a range of 15 km from the radar.

A UAV-based synthetic aperture radar could provide multiaspect information on stationary targets, sufficient to permit target detection and classification. Algorithms for automatic target cueing and recognition (ATC/ATR) are projected to improve considerably. The false-alarm density at a 50 percent detection probability may decrease by one or two orders of magnitude from the current performance of one per 10 km^2. High-performance ATC/ATR algorithms, running on high-speed computers, can provide real-time detection of targets at surveillance rates (measured in square kilometers per second) that would overwhelm a human's imaging and decision capabilities. The underlying technologies for these advances will include neural nets, statistical pattern recognition, and model-based vision.

UAV-based radars could also be used for low-altitude air defense, overcoming the difficulties with terrain and foliage masking that hamper ground-based air defense systems.

Geographically dispersed *networks of acoustic sensor arrays* can be used to detect, locate, and recognize aircraft, weapons that are firing, and ground vehicles. While single arrays can provide directional cueing, networks of arrays can locate weapons and track aircraft. Networked acoustic arrays would provide passive battlefield surveillance for a variety of targets and cueing for active sensor systems. The electronics needed for these networks include noise suppression for the arrays, small data processors deployed with each array, low-data-rate communications from all arrays in the network, and processors to apply interpretation algorithms to the data collected from the network. The Technology Group forecasts that major capabilities of this type might be achieved within 5 years (Table 3-1).

Superconducting quantum interference device (SQUID) magnetometers are sensitive enough to detect the magnetic field perturbation generated by a moving tank at near range (Figure 3-12). However, ambient temporal variations in the earth's magnetic field are four to five orders of magnitude larger, so background noise is likely to obscure the tank's signature. Whether advanced signal processing could distinguish tank signatures is a question requiring further field measurements and research.

Ground-based radar systems constitute the principal surveillance-

TABLE 3-1 Current and Projected Capabilities of Acoustic Array Sensor Networks

Parameter	Current Capability (experimental 5-m array)	Projected Capability (smaller 2-m array)
Detection range of 5 to 20 km (10 km avg.)	Low wind and quiet background noise	High winds and high background noise (battlefield conditions)
Direction finding of 2° to 3° accuracy and 15° resolution	3 loudest targets under same "quiet" conditions	3 to 5 loudest airborne targets and several loudest weapons under battlefield conditions
Target location within 50 to 1,000 m	Depending on network geometry and source motion	Depending on network geometry and source motion
Multitarget location weapon	1 airborne target per array; unknown for transients (weapons) and ground vehicles	1 to 3 airborne targets per array; several weapon firings per second per array
Recognition	Single helicopters in quiet background	Helicopters in multi-target, noisy environment; recognition and aid for other aircraft

and-tracking sensors for surface-to-air missile systems. Current systems, including HAWK and Patriot, are severely strained by newer and potential threats: tactical ballistic missiles, low-observable aircraft, cruise missiles, and modern electronic countermeasures. Advanced technologies for improving these systems are in development but were not detailed by the Technology Group.

The Technology Group summarized the performance parameters and technology requirements for *space-based radar platforms for surveillance and target recognition*. This application would require *lightweight phased array radar antennas*. The Technology Group assessed the advantages of corporate-fed phased array antennas for this purpose and for other applications, such as long-range, low-observable airborne radar. The transmit/receive functions of array antennas are an excellent application of the MMIC technology described above.

Summary Findings: High-Impact Electronic Technologies

Terahertz electronic devices are those that can operate at frequencies up to 10^{12} Hz. They will be needed to amplify and process analog signals with frequencies extending to this limit; they also will provide the basis for digital logic that can be switched within time intervals on the order of a picosecond (10^{-12} s). Terahertz devices will be used as the fundamental components in advanced radar, communications, electronic intercept, and weapon guidance/seeking systems. The electronics technologies that are potential candidates for terahertz performance include devices based on compound semiconductors (e.g., GaAs and InP), superconductive devices, vacuum microdevices, and optoelectronic devices.

Teraflop computers are high-speed computers capable of performing 10^{12} floating point operations per second. Unless logic devices with switching speeds of less than 10^{-14} s (two orders of magnitude faster than the terahertz devices discussed above) can be implemented, which

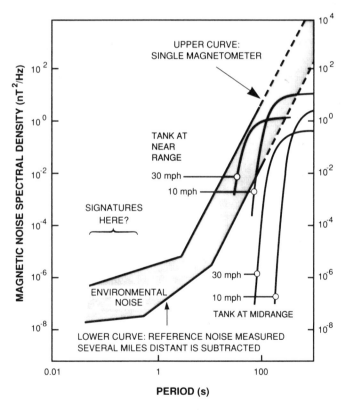

FIGURE 3-12 Comparison of tank magnetic signature with environmental magnetic noise.

seems unlikely, a teraflop computer will require a large number of slower processors operating in parallel. For example, a single processor based on terahertz devices may be able to achieve a computing power of 10^{10} operations per second. So a teraflop computer based on terahertz devices would require about 100 such processors operating in parallel.

High-resolution imaging radar sensors will use the terahertz devices and teraflop processors described above for radar sensor suites capable of finding and recognizing targets at long range. Targets could be either stationary or moving; fixed targets will be the more difficult detection-and-recognition task, which will probably require the teraflop processing capability. The ability to locate and identify surface targets with precision at long range, combined with advanced communications, navigation, and command systems, will enable such targets to be attacked successfully from distances well beyond the range of enemy weapons.

OPTICS, PHOTONICS, AND DIRECTED ENERGY

TFA Scope

This Technology Group assessed the following areas:

- *Optical sensor and display technologies* receive optical radiation and interpret it for imaging displays to the user. The technological advances in this area include laser radar; multidomain sensors; sensor fusion; infrared search, track, and identification; focal planes with massively parallel processing; and helmet-mounted or similar heads-up displays.
- *Photonics and optoelectronic technology.* Photonics comprises the science and technology to use photons to transmit, store, or process information. Optoelectronic technology couples electronic data-processing elements with optical elements. This area includes fiber optics, diode laser arrays, optoelectronic integrated circuits, optical neural networks, acousto-optical signal processing, and various other technologies that process optically transmitted information.
- *Directed energy devices* are intended to generate highly concentrated radiation as a means of directing a high level of energy—which may be very short in duration—on a small target area. The radiation may be from the optical portion of the electromagnetic spectrum (as in lasers), from the radio frequency portion (microwaves), or from accelerated charged particles. Directed energy technology was considered by the Power and Propulsion Technology Group as well as the Optics, Photonics, and Directed Energy Technology Group.

Technology Findings

General Findings

Optical sensor technology and photonics provide basic building blocks for advanced integrated sensors and high-speed processors. Directed energy devices provide the long-range, speed-of-light capability to degrade or destroy hostile smart systems.

Essential for advanced system design in these technology areas is a computer-aided design environment that allows the integration of detailed information on sensors, processors, and the basic properties of their components. In addition to the initial design and development effort, such an environment could be used interactively to respond to evolving threats.

Optical Sensor and Display Technologies

Laser radar provides high-resolution target imaging, target discrimination, and detection of low-observable targets. The current technology includes systems based on carbon dioxide lasers, solid state lasers that use diode-pumped neodymium, and titanium sapphire lasers.

Solid state laser technology is being extended to an average power of several hundred watts, which will enable laser radars to have very long ranges (depending on the wavelength and atmospheric attenuation). Further development of both carbon dioxide and solid state laser technology should provide the peak and average power needed for various laser radar applications, while the size of the system will decrease significantly.

Multidomain smart sensors will combine a laser radar with one or more other sensor systems. A laser radar working with a wide-area surveillance sensor, such as microwave radar or a passive infrared search-and-track (IRST) sensor, enhances target detection and identification while reducing false alarms from clutter. Because the system can be configured so that the sensor components share the same physical optics, information across domains can be fused at the pixel level. This can provide a multidimensional information space for subsequent sensor fusion processes. The richness of this information can allow a human observer to detect targets in motion and stationary targets—even those concealed by camouflage or ground cover. One such system, for use on tanks and air defense weapons, combines a rangefinder and front-looking infrared laser radar.

Passive IRST systems can be used with laser radars for wide-area searches capable of detecting low-flying aircraft against terrain and clouds. Passive IRST has the added advantage of being covert; it sends out no signal beam that can be targeted by enemy counter-measures. By operating in two bands simultaneously, passive IRST can make stealthy or camouflaged air vehicles more detectable.

A multidomain sensor system with laser radar and sensor fusion at either the pixel level or the image level will be part of an auto-matic target recognition system for detecting and classifying aircraft and missile threats on the tactical battlefield. Airborne systems cur-rently under test for tactical target detection and identification use carbon dioxide and GaAs laser radars in combination with passive sensors in the visible and 8- to 12-μm region, plus 85-GHz millimeter-wave radar.

Another application of laser technology for multidomain smart sensors is differential absorption LIDAR, or DIAL. (LIDAR stands for light detection and ranging.) DIAL can be used to detect specific chemicals in atmospheric emissions by their absorption of light from one laser beam of a dual-beam system. Current systems are being developed to detect volatile solvents used in clandestine chemical-processing operations (drug processing in particular). The technol-ogy is extendable to the detection of other military targets, includ-ing CTBW production facilities, vehicles hidden in trees with their engines idling, fuel dumps, and perhaps ammunition dumps.

Further experience is needed in combining laser radar with pas-sive IRST in a package suitable for Army applications. If a first-generation system can be field-evaluated within the next 5 years, full production of the system should be possible within 15 years. A passive IRST wide-area search system combined with laser radar for ranging and identification would have a major effect on low-altitude surveillance and defense against air targets. The variety of such tar-gets would include conventional and stealthy manned aircraft, cruise missiles, UAVs, and tactical missiles.

Only an integrated approach to *sensor fusion* can satisfy the demands of the future battlefield for rapid integration and interro-gation of signals from a multisensor suite (Figure 3-13). Successful response to incoming missiles, aircraft, smart weapons, and satellites will require completely autonomous target detection, recognition, and acquisition. The time requirement may not permit a man in the loop, so the system must provide 100 percent target validation. These time and reliability requirements necessitate the use of multidomain sensors and automatic processing of their images.

With respect to sensor fusion technology, a new concept is the *inte-grated sensor*. A high-capacity, optical-domain parallel processor—prob-ably of a neural net design—would be directly interfaced to the high-

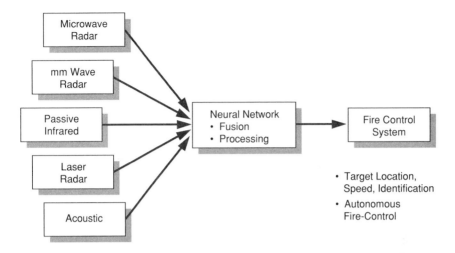

FIGURE 3-13 General concept of an integrated sensor fusion device.

density focal plane of a multidomain sensor optics package. The output from the processor would feed directly to a fire control system.

Focal plane arrays are currently available for the infrared region, but integration into a monolithic, switchable structure has not yet been achieved. Optical parallel processing is under development. Signal processing from multiple acoustic receivers, whose output might go to the fusion unit through a secure fiber-optic channel, is another development requirement.

Smart focal planes are another concept for the rapid processing of data from sensor optics (Figure 3-14). An array of small-area detectors will share space on the focal plane with processing circuitry. An array of microlenses will direct the incoming radiation to the detector array. The focal plane image will be read out in a massively parallel manner to a sequence of optoelectronic processing planes beneath the focal plane. This parallel-processed readout from the focal plane will avoid the current bottleneck of serial readout and serial processing in conventional serial computers.

As conceived, the smart focal plane technology would allow image acquisition and processing rates greater than 5,000 frames per second. The output information will be highly processed already, so communication bandwidths can be reduced. The potential size reduction from current serial processing technology could yield advanced capabilities in a package small enough for use in smart missiles. However, much of the technology to implement the smart

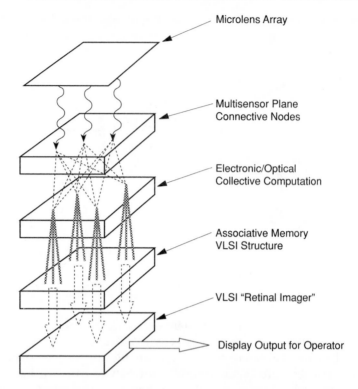

FIGURE 3-14 Concept for a future smart focal plane device.

focal plane concept remains to be developed. Suitable algorithms and processor architectures for optical processing of images are still in the research stage.

For infrared scanners, focal plane arrays of detectors based on *Schottky-barrier materials* can improve the photon collection efficiency of the entire sensor by five orders of magnitude, compared with conventional infrared scanners. Although Schottky-barrier materials have a lower quantum efficiency than other solid state detectors, a focal plane array in a staring format compensates for that disadvantage by using a large number of detectors. Arrays of $10,000 \times 10,000$ detectors should be available in the next two decades.

Schottky-barrier technology covers all optical wavelengths of interest to the tactical battlefield—visible, near-infrared, and mid-infrared. There are many options for combining them in multi-spectral arrays or tuning for a particular region. Arrays for the near-infrared region can use skyglow and thermal emission in the 1- to 2-μm range for night vision.

Smart sensors will also be applicable to the soldier's personal gear, as in the *smart helmet* concept (Figure 3-15). Future battlefields will require enhanced awareness by the field soldier, in response to increased use of camouflage and stealth techniques. Eye protection will be needed against antipersonnel lasers. The smart helmet incorporates advanced night vision sensors, sensor processing, and commu-

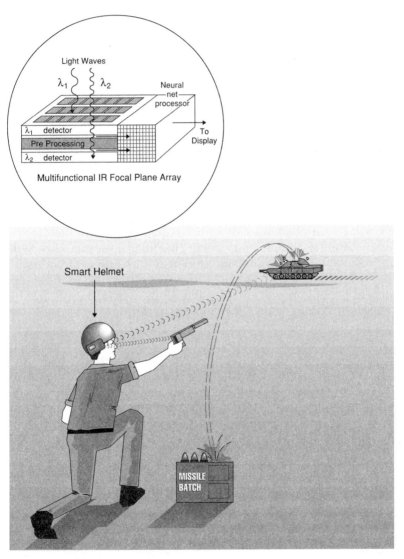

FIGURE 3-15 Smart sensor (infrared focal plane array) with Army applications for smart helmet.

nications with eye protection, because the information seen by the eye will always be an indirect display.

Notable advances in *display technology* will include the integration of monolithic drive circuitry with lightweight, low-power display arrays. Displays will range from personal "eyepiece" viewers and helmet-mounted heads-up displays to large, multiviewer display screens. The direction of development in advanced displays will be toward allowing a human operator to have true telepresence in environments that are too dangerous or are physically inaccessible. Lightweight displays with a wide field of view will have major military applications.

Photonics and Optoelectronic Technologies

Photonic approaches to communications, such as fiber-optic cables, offer several advantages over electronic systems. They are relatively immune to electromagnetic interference and provide very large bandwidths (in the terahertz range). Computing applications of photonic systems can have higher clock rates and large-scale parallel processing.

For military *communications network* applications, a fiber-optic network will be supplemented by radio frequency links to mobile nodes: sensors, satellite relays, and users. The fiber-optic network will be more resistant to jamming, interference, and interception than the more vulnerable radio links. Military applications of fiber optics currently provide rapidly deployable links over distances from tens of meters to kilometers. Time-division multiplexing is used, which limits the bandwidth and compromises the robustness and flexibility of the network. Future military applications will combine wavelength-division multiplexing with time-division techniques, providing a combined peak network capacity in the range of 10 Gbit/s or more and servicing hundreds of users.

Fiber optics may also allow *close integration of wide-bandwidth sensors* with ground operations. In addition to advanced C^{3}I sensors, this capability will also enable *telepresence* by passing high volumes of sensory data between the remote platform and the ground operator, via a connecting optical fiber.[2] Because the high-powered signal

---

[2] Members of the STAR Committee expressed reservations concerning the range of practicality for battlefield applications such as wide-area networks or tethers to high-performance ground or air platforms and long-range missiles that communicate to a ground-based controller via a fiber-optic link.

processing and computing capabilities needed for data reduction, interpretation, and display can be located at the ground controller's location, the remote platform can be much smaller, less expensive, and therefore more expendable. Among the possible applications are (1) advanced fiber-optic-guided missiles, (2) airborne surveillance platforms with multidomain smart sensors but minimal onboard signal processing, and (3) tele-operated ground vehicles for both reconnaissance and weapon delivery.

Guided optical-wave sensors are an area of fiber-optic technology in which changes in the amplitude or phase of optical waves in the fiber are used to sense vibration (acoustic or seismic sensors), temperature or pressure changes, rotation (gyroscopic sensors), and even electrical or magnetic fields. A notable current effort in this area is the fiber-optic gyroscope. Although it is less sensitive than mechanical or laser gyroscopes, it offers compactness, robustness, and low cost—qualities that suit it to a number of missile applications. This technology is just emerging.

Since the mid-1980s, important advances have occurred in *solid state laser technology*, largely through research supported by the Department of Energy, DARPA, SDIO, and the service laboratories. Their all-solid-state design provides the advantages of high reliability and low maintenance. Mass production techniques for solid state materials and for laser array pumps promise low cost as well. At present, designs with longitudinal pumping give the highest efficiencies, but transverse pumping of solid state laser slabs by two-dimensional diode laser arrays is better suited for higher power levels, albeit at modest efficiencies.

For example, the projected capability of a neodymium laser demonstrator, due by 1993, is 300-W average power, about 10 percent efficiency, and a lifetime of greater than 10^9 shots. Various wavelength conversion techniques will enable this demonstrator or similar devices to be wavelength-selectable from the visible to the mid-infrared at greater than 100 W. This average power level exceeds the minimum required for many space-based and tactical applications.

This solid state laser technology will provide eye-safe laser rangefinders and target designators that are more reliable than those now fielded. In addition, however, it will lead to new applications, such as the laser radars described earlier, active optical countermeasures (antisensor lasers), and high-bandwidth laser communication from satellites to theater and battlefield commanders.

A related area with recent advances is *diode laser and laser array technology*. While individual diode lasers are limited in power, coherent arrays will yield 10 W or more of output at greater than 40 percent efficiency. Scaling to over 100 W average power may be feasible, with power densities exceeding 100 W/cm². Varying the semiconductor material will enable these arrays to operate in the visible region, at eye-safe wavelengths above 1.4 μm, and even in the 2- to 5-μm range that is used for laser-activated proximity fusing.

Optoelectronic integrated circuits combine electronic and optical microcomponents on a single semiconductor chip (Figure 3-16). The purpose of the chip may be to provide an information interface between the two technologies or to create a functional hybrid device. Commercial applications are driving the rapid development of this just-emerging integration of the two technologies. Within 15 years, optoelectronics will be mature for communications and computer interconnect applications. In 30 years, it will have a wide range of applications built on hybrid functionality, such as massively parallel optical processing and wavelength multiplexing.

The best-characterized materials for optoelectronics are GaAs and other semiconductors formed by combining Group III and Group V elements (III-V semiconductors). Ferro-electric liquid crystals are another possibility, particularly for light-modulating applications. Lithium niobate is currently the material of choice for volume holographic storage and interconnects.

A key point is that all the semiconductor optoelectronic technologies, even of the III-V semiconductors, are still very immature compared with silicon technology. A substantial and sustained investment will be needed for this technology to mature. In the near term, optoelectronic-processing applications will mostly use arrays in which the logic function is performed by electronic components, while the optical components provide the mechanism for highly parallel interconnections. In terms of combined speed, low power, and high spatial density, optoelectronic arrays based on III-V semiconductors will be difficult to surpass in the long term.

Neural networks constitute another information-processing technology in which photonics will play an increasing role. By analogy with biological neural systems, a neural network contains two types of processing elements: synapses and neurons. A synapse performs an operation on its single input; a neuron receives inputs from multiple synapses and combines them in some nonlinear way to produce an output. Although photonic or optoelectronic implementations of these elements are at present less developed than electronic alternatives, they offer the potential for far larger numbers of synapses and neurons per component. In addition, photonic elements can support more flexible connectivity patterns, including some that appear essential for neural net architectures to perform vision and image-

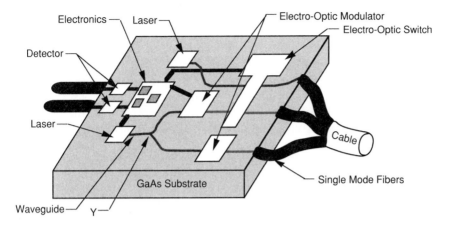

FIGURE 3-16 A simple optoelectronic integrated circuit.

processing tasks. In the long term, the Technology Group forecasts that optical neural networks will be used for real-time automatic target recognition based on multidomain sensor inputs, for speech understanding (i.e., word and pattern recognition), and for complex signal processing.

The TFA includes a special chapter on existing R&D projects on optical neural networks.

Acousto-optics uses the crystal vibrational modes of a Bragg cell to encode or decode information carried in the modulation of light beams. The potential information-carrying capacity of modulated light and optical processors can be illustrated by an analog acousto-optical device called a time-integrating acousto-optical correlator. Current versions of this device can process the equivalent of 10^{13} operations per second, which is several orders of magnitude more than existing electronic devices and a factor of 10 above the goal of "terahertz" devices. In addition to this potential for higher information throughput, optical processor architectures will require less power and will be smaller and weigh less than digital electronic processors.

Acousto-optics is also applicable to advanced sensors. The characteristics that can be "read" from a signal intercepted from an emitting platform can be used to identify the specific signal type and determine the location or velocity of the platform. There is no limitation on the wavelength of the electromagnetic radiation that can be processed in this way.

Optical techniques with lasers can be used to control information carried in the amplitude, frequency, and phase modulations of *micro-*

wave radiation. In addition, optical fibers make excellent waveguides for distributing information-carrying microwaves; the available bandwidth can be on the order of hundreds of gigahertz. This technology appears promising for control of phased arrays, control of remote antennas, microwave communications requiring extremely high data rates, and secure communications. The Technology Group forecasts that within 15 years microwave and optical circuitry will be integrated on a single chip. In 30 years, sophisticated optical computing will be used for various adaptive antenna functions, such as beam shaping, null steering, and side-lobe suppression. Many microwave frequencies will be multiplexed over one fiber-optic network.

In *adaptive optics,* a wavefront sensor is used to measure aberrations in an incoming light signal. This information controls a deformable optical element that adjusts to compensate for the optical aberrations, which would otherwise limit the performance of the optoelectronic system. The advanced techniques for adaptive optics use nonlinear optical materials that perform both the sensing and the compensation functions.

Aberrations caused by optical system imperfections or the atmosphere result in substantial signal degradation or loss of laser coherence in nearly all present optical systems. Adaptive optics will become an essential part of future systems; they will substantially increase the operating range and improve the resolution of laser systems. For military applications, adaptive optics can improve performance of many optoelectronic systems, including antisensor lasers, passive battlefield imaging, active or passive space object imaging, auto-tracking, and optical jammers. Adaptive optics can also improve the projection of laser power from directed energy weapons by correcting for atmospheric aberrations between the beam source and relay mirrors.

Sophisticated countermeasures to laser antisensor threats can make use of *applied nonlinear optics.* This technology will also be important in implementing components required for all-optical processing and computing. The Technology Group forecasts its use for passive laser protection within 15 years and for advanced optical processing components in 30 years.

Binary optics is a technology for creating diffractive optical devices on a substrate by use of lithography and micromachining. It gives the optical circuit designer the capability to create novel elements as well as alternatives to more conventional refractive elements. Binary optics methodology builds on VLSI circuit technology; both are well suited to computerized circuit design and manufacturing. Low-cost mass production of binary optic designs is possible through replication, embossing, or molding of subassemblies. In addition, the potential of diffraction devices to compensate for aberrations pushes the range of optical design further into the deep infrared and ultraviolet regions.

The transfer of binary optics technology to U.S. industry began only in 1988. Already, more than 30 optics and aerospace companies have acquired the knowledge and capability to produce binary optics. One important near-term Army application for binary optics is to correct chromatic aberration in infrared imaging systems. Binary optics also has the potential to simplify the production of optical systems for military applications. It should also make those systems cheaper, lighter, and lower in power consumption.

Directed Energy Devices

In a technical sense, even laser devices that are used for information functions (laser radar, rangefinders, target detectors) can be considered directed energy devices. However, as used here the term applies to energy beam technology primarily concerned with delivering a high-energy flux on a target.[3]

The Technology Group described the technologies needed for a conceptual ground-based laser antisatellite system. The directed energy weapon in this system would either be a free-electron or chemical laser, complemented with adaptive optics.

The *free-electron laser* (FEL) uses a high-energy accelerator to create an intense stream of electrons. The stream traverses a series of alternating magnetic fields, which causes them to emit coherent electromagnetic radiation at a wavelength tunable by the electron's energy and the magnetic field strength. The entire beam-generating process occurs in a high vacuum, which limits self-distortion found in crystal or gas lasers. The distinctive advantages of this high-energy laser include efficient production (greater than 25 percent efficiency) of high average-power output; broad, continuous tunability over a wide frequency range (in theory, at least, from long-wavelength microwave to short-wavelength x ray); excellent beam quality; and generation by electrical power, which simplifies logistical support.

The Army is pursuing two FEL approaches under SDIO funding. Both have demonstrated sufficient electron beam brightness for operation at high power and have been operated at a variety of wavelengths (although not at x-ray wavelengths). Peak output power in excess of 10

---

[3]The Propulsion and Power Technology Group considered directed energy weapons in their TFA as well. That discussion covers the same areas but in somewhat more detail than in the Optics, Photonics, and Directed Energy TFA.

MW has been demonstrated, with average power for short durations in the range of tens of kilowatts, but excellent beam quality at high average power is yet to be demonstrated. Major testing is under way.

By 2020, ground-based—and possibly space-based—FEL systems might conceivably be able to intercept and destroy missiles during their boost phase. (A ground-based system would use space-based relay mirrors to reflect the beam onto targets.) To achieve this goal, difficult technical breakthroughs in beam generation and steering are required. For tactical applications, a high-power microwave beam, using an FEL source, should be available with multimegawatt power.

High-voltage, short-pulse electron beam accelerators can be used to drive conventional microwave sources (magnetrons, klystrons, backward wave oscillators) to create an intense, narrow-band, pulsed *radio frequency energy beam*. Peak power levels can be as high as several gigawatts, with energies per pulse greater than 200 J. A newer technology uses a solid state switch to produce a high-power wideband radio frequency beam (also called a video pulse), which can operate as a repetitive pulse (repeating with a frequency of 10 to 100 Hz).

Continued progress with these high-power, high-energy radio frequency sources will enable a new class of weapons, in which mission kill is accomplished by burning out electronic components or detonating electro-explosive devices in the target. Potential targets include smart munitions, antiradiation missiles, mines, aircraft, radar and infrared guided missiles, communications nodes, and UAVs. The Technology Group also summarized the projected capabilities of systems now under development.

FEL, high-power microwave, and charged-particle beam weapons all depend on the development of a *compact accelerator*. The basic concept is to alter the linear transport geometry of the traditional linear induction accelerator into a spiral or circular configuration. This would allow the same accelerator module to act repeatedly on a circling swarm of charged particles, until they reach the desired velocity. Compact accelerator development is being pursued by the Naval Research Laboratory and in two projects currently supported by DARPA.

BIOTECHNOLOGY AND BIOCHEMISTRY

TFA Scope

The Technology Group on Biotechnology and Biochemistry characterized "biotechnology" as the application of scientific principles for clinical and industrial uses of biological systems to produce goods and services. Living organisms or their parts are used to make or modify products or to develop organisms for specific purposes. Contributing technologies include molecular and cellular manipulation; enzyme definition, design, and production; and microbial techniques for growth and fermentation. The technologies assessed by this Group were divided into the following six categories:

- *Gene technologies* include the methods to "touch the genome" and modify it. (The genome is the cellular site of genetic material, which carries the information for biological inheritance.) The techniques include gene replication, splicing, modification, regulation, transportation, and expression.
- *Biomolecular engineering* is the technology to design and produce biomolecules (structural proteins, enzymes, etc.) with specific, tailorable properties.
- *Bioproduction technologies* use living cells to manufacture products in usable quantities. These methods of biosynthesis can range from fermentation to solid state molecular synthesis, multistage bioreactors, and methods still evolving.
- *Targeted delivery systems* are composites of materials that are designed to concentrate the active agent(s) in the composite at specific sites in the body where its activity is desired.
- *Biocoupling* is the linkage of biomolecules or biomolecular complexes to electronic, photonic, or mechanical systems. For example, highly sensitive and selective detector molecules (biosensors) would be bound to microelectronic (or optical) circuitry to produce a system able to detect a single molecule of a CTBW agent.
- *Bionics* aims at methods of directly connecting the human neural system to electronic or mechanical systems, such that the nonhuman system functions in ways similar to human limbs or organs.

Technology Findings

General Findings

The technologies in the scope of this TFA are based on a wide range of scientific disciplines (molecular biology and biochemistry, physical and organic chemistry, medicine, manufacturing process technologies, and electronics). To exploit the potential of biotechnology for Army-specific applications, the Army will need to assemble multidisciplinary research teams with competence in physics, chemistry, biology, medicine, and engineering, rather than segregate staff by discipline, as traditionally done.

Collectively, the technologies are the newest, least mature of the STAR technology areas. They are expanding rapidly in terms of discoveries, applications, and inventions. Also changing rapidly are perceptions of their importance. The possibilities are high for amplification of results from relatively small increments in investment. The Technology Group believes that a stable funding base for biotechnology is essential to provide the continuity of research and application development required for its military potential to be realized.

The Technology Group believes that success in achieving biotechnology goals depends on program management with a fundamental appreciation of advanced molecular biology, especially nucleic acid and protein chemistry, immunology, and infectious disease (especially vector biology). An understanding of process engineering is also key to success. Program management for Army biotechnology programs should reside in domains such as the U.S. Army Materiel Command, the Medical Research and Development Command, and the Chemical Research Development and Engineering Command.

In some areas of direct interest to Army applications, foreign biotechnology is ahead of U.S. academic and nonmedical private sector efforts. The Technology Group concluded that the Army is well poised to use work from important foreign laboratories, because the Army Materiel Command has nascent but important programs in many of them. Various administrative, contractual, and legal restrictions must, however, be overcome to realize this potential. A specific recommendation is to establish with our allies more joint military technical working groups. The management of such working groups, or other forms of joint research teaming, should include currently active researchers as well as senior technical managers. The Technology Group believes the future of this area can best be foreseen by those at the research bench.

The Technology Group foresees CTBW as a growing military threat. Potential adversaries will continue to find the use, or merely the threat, of CTBW as an inexpensive weapon of aggression and defense from retaliation by the world community. The Group believes deterrence can best be achieved by a system of countermeasures, which would collectively eliminate the efficacy of the CTBW threat to Army contingency operations. The discussion in Chapter 2 of CTBW countermeasures, including (1) detection and identification, (2) physical protection, (3) medical prophylaxis and therapy, and (4) decontamination, was drawn primarily from an appendix in this TFA on CTBW. The Technology group forecasts that biotechnology will prove pivotal in all four of these countermeasure categories.

Today's successes in biotechnology cluster in the areas of medicine and pharmaceuticals, agriculture, and bioproduction of specialty "natural-product" chemicals such as sweeteners and solvents. The Technology Group foresees future capabilities extending to large-scale bioproduction from generic feedstocks, design and synthesis of novel biomaterials; coupling of biomolecules with electronic, optical, and mechanical devices; selective improvement and modification of life forms; and environmental decontamination (bioremediation).

For the kinds of advanced capabilities needed by the future Army, biotechnology offers important advantages when compared with traditionally engineered and manufactured systems:

• Biological systems perform complex, repetitive syntheses with few side products and few errors, compared with traditional chemical production methods. They are therefore well suited to *routine, reliable production of complex substances in pure form.*

• Bioproduction can give these complex substances extremely specific recognition capabilities, making them ideal for *selective detection or site-specific activity.*

• Biochemical reactions typically occur under milder conditions than analogous industrial chemical processes, so bioproduction can be less expensive, require lower energy inputs, have less critical operating conditions, and require simpler apparatus.

• The cost of routine production of biotechnology products should usually be less than for alternatives produced by traditional processes. In most cases of biomolecular-engineered products, there will be no comparable alternative producible by nonbiological processes.

• Biosystems are "engineered" at a molecular level, so such systems (a white blood cell, a microorganism, an eyeball, or a brain) are very compact relative to a traditionally manufactured electrical, optical, or mechanical system with similar functionality. To the extent that this engineering uses "components" already developed by nature, a large part of the initial R&D cost has already been paid.

• The most important "system" in the future Army will continue to be the human soldier. Because the soldier is a biological system, biotechnology offers unique potential for enhancing the performance of this most complex, critical, and costly of the Army's systems.

A major obstacle to achieving biotechnological advances is the lack of adequately skilled personnel. The Technology Group addressed this issue in detail in a special chapter.

Gene Technologies

Recombinant DNA techniques can now be used to transfer the characteristic of one or several specific genes to a different cell or organism. As knowledge of specific genes and the mechanisms by which they interact increases, the techniques of recombinant DNA, cell fusion, and gene splicing will permit the transfer of multigene, complex characteristics into cells and organisms.

Cell fusion, or hybridoma, technology involves the fusion of two cells, each with desired characteristics. For instance, a cell that produces a specific antibody can be fused with a cell easily grown in cell cultures. The hybrid cell retains the ability to produce the desired antibody but can be easily cultured in quantity. A recently announced hybridoma technique allows the production of monoclonal antibodies in days, rather than the months formerly required. Similar "quantum-leap" advances will continue for at least the next three decades.

Gene technologies enable production of new substances, or even new organisms, with applications to medical and nonmedical interests of the Army, such as these:

• substances for discrete recognition of an organism (including identification of individual persons) or a substance (DNA probes, receptors and antibodies for specific molecular conformations);
• diagnostics for disease and CTBW threat detection;
• new or altered materials, with improved structural, functional, or renewable characteristics, produced by genetically altered biological systems;
• medicinal drugs and therapeutic agents;
• vaccines and multivalent vaccine delivery systems;
• physiologically active compounds that modify biological response;
• artificial body fluids and prosthetic materials;
• new foods and food production processes;
• decontamination, detoxification, and bioremediation processes;
• new or improved materials for adsorbing or neutralizing hazards and for purifying water, food, or production feedstocks.

Gene technologies will be used in all seven of the high-payoff opportunities described below.

Biomolecular Engineering

Our current ability to relate the structure of biomolecules to their function is limited for all but the smallest of these molecules. There are still many surprises and predictive failures, even in areas where predictive methods are most advanced. At present we lack the ability to design *de novo* a biomolecule for a reasonably complex function, such as radar nonreflectivity. However, the scientific disciplines to pursue such a capability do exist.

Progress in biomolecular engineering will depend on advances in two contributing areas: (1) prediction of the biomolecular structures required to achieve a desired function and (2) methods to design, construct, and produce molecules or composites that meet specific functional requirements. The multidisciplinary research teams needed for this work must combine expertise in structure-function physical chemistry; physical biochemistry; computational methods for simulation, modeling, and display of biomolecules; analytical methods for determining the detailed structure of biomolecules; biophysics and chemistry of molecular biopolymer synthesis; and the biochemistry and molecular genetics of the genome.

Of the seven high-payoff opportunities identified by the Technology Group (see below), biomolecular engineering will be applicable to five: deployable bioproduction of military supplies, biosensor systems, novel materials, extended human performance, and antimateriel products.

Bioproduction Technologies

The bioproduction techniques and resources already available include bioreactors; cell culture and fermentation techniques; cell growth media and factors; established cell lines for mammalian, insect, bacterial, yeast, and algal cells; cell harvesting and processing techniques; chemical coupling techniques and processes for immobilizing (fixing) cells and proteins; and techniques for purification and isolation, such as affinity chromatography.

Further development of fermentation and cell culture techniques, cell lines, and bioreactors will be particularly important for efficient large-scale production. Bioproduction methods also need to be scaled up from laboratory size to industrial production scales.

Affinity chromatography is based on the covalent coupling of affinity ligands, enzymes, and other biomolecules with specific recognition characteristics to inert, solid support materials. The resulting technology will enable rapid, efficient purification and processing of ultrapure materials on a large scale. In one type of purification (monoclonal antibodies), the older technology of column chromatography had a process yield of only 40 to 60 percent, gave a product that was 95 percent pure, and required 2 to 3 days. The new method based on membrane affinity can process the same amount of material in 1 hour, giving a 90 to 96 percent yield and a product that is 99 percent pure.

Bioproduction technology will be applicable to five of the Technology Group's selected high-payoff opportunities: deployable bioproduction of military supplies, enhanced immunocompetence, novel materials, in-field medical diagnosis and treatment, and anti-materiel products.

Targeted Delivery Systems

In a targeted delivery system, an active substance is encapsulated in a membrane or a matrix that permits controlled release when the capsule system reaches its intended site of action. The release may be slow, by diffusion out of the encapsulating material, or triggered by dissolution of the capsule. Thus, these systems permit the use of biosubstances that would otherwise be inactivated or degraded before they could be effective for their intended purpose.

New microencapsulation technology, using biomaterials that are biocompatible and biodegradable, will protect sensitive active substances from degradation or inactivation by light, chemical, or biological stresses. In medical applications, drugs, vaccines, peptides, and proteins will be administered with microencapsulation systems now under development. In nonmedical applications, field-deployable, stable capsule systems will be useful for intelligent biosensors, decontamination systems, and biocamouflage systems for signature suppression.

Among the potential applications of interest to the Army are drug and vaccine delivery systems for prophylaxis or treatment of infectious diseases or CTBW agents, energy-rich or performance-enhancing foods and supplements, decontamination methods, deployable purification kits, and regeneration or replacement of tissues and organs.

The Technology Group projects advances in this area that will produce self-regulating delivery systems. Specific triggering mechanisms for release of the active substance will be developed, such as triggers by pH, ionic strength, specific receptor/ligand binding, or specific frequencies of electromagnetic radiation.

Of the high-payoff opportunities for biotechnology, targeted delivery systems could play a role in enhanced immunocompetence, novel materials, in-field medical diagnosis and treatment, and anti-materiel products.

Biocoupling

For near-term biosensor applications and longer-term bioelectronics, it is necessary to develop techniques to couple the biocapture and recognition event (the response of a biomolecule to its target molecule or energy form) to the means for amplifying, transducing, and communicating that information into an electronic, optical, or mechanical signal. Development of antibody or bioreceptor molecules as biosensors is in progress. The coupling technology is less advanced, receives less attention, and will be more difficult.

Bioelectronics refers to the use of biomolecules or biosensor systems within an electronic data-processing system—for example, a "microchip" integrated circuit that incorporates biosensor elements into a computer memory "biochip." The development of this technology depends not only on biocoupling advances but also on biomolecular elements with a binary signal response.

As with biomolecular engineering, to reap the potential of biocoupling technology will require multidisciplinary teams competent in many specialties, including molecular genetics, receptor physiology, and pharmacology; physical chemistry of macromolecules; the physics and chemistry of signal trapping and recognition; engineering adaptation of unit-event signals into systems with integrated outputs; and engineering to adapt the environment required by the biosensor to the sampled environment.

The Technology Group recommends that biocoupling be pursued in parallel with biosensor development, because biocoupling methods may determine which biomolecular mechanisms are feasible as biosensors within the larger system to which they are coupled.

The projected applications for biocoupling include the following areas of interest to the Army:

• deployable remote detection and analysis systems (with telemetry) to assess the presence and status of hostile troops and equipment, disease and CTBW threats, or environmental parameters;
• rapid diagnosis and identification of disease and CTBW threats in the field;
• terrain and perimeter monitoring;
• monitoring of critical personnel performance;
• performance modification (see also bionics biotechnology); and
• bioelectronics.

Bionics

A successful bionic device replicates both qualitatively and semiquantitatively the function of the living physiological system it imitates. The technology has been most successful in copying neural systems that couple an environmental signal to receptors that act as high-gain analog-to-digital converters and otherwise process the signal as information.

Passive bionics, in which properties of a biological material are emulated for a single objective, is a maturing technology already being exploited by the Army. Potential developments include fibers for personal armor, nonlinear optical polymers for eye protection from lasers, and artificial sense organs for robotics applications. Bioelastomers, antifoulants, biolubricants, and bioadhesives are under study in important current programs. The Technology Group anticipates some striking successes in this area but no surprises.

The alternative to single-property, passive bionics is multicomponent cybernetic systems that model the neurally modulated "smart" behavior of animals. Such systems could be possible in 30 years. The Technology Group expects that progress in this area will depend on the development of high-density neural nets with advanced artificial intelligence capabilities and on advances in biocoupling.

Highest-Payoff Opportunities

The Biotechnology and Biochemistry Technology Group identified seven areas of biotechnological application as having the highest payoff for meeting long-term Army needs. It then matched these application areas to the advanced systems concepts of the STAR system panels. On the basis of the resulting matrix, the Technology Group identified one or more key products that could be produced within the next 30 years in each of the seven areas.

For each high-payoff area, the Technology Group set up a *road map*, or R&D time line. These road maps show the timing of required Army investments in the enabling technologies, important milestones in the development process, and an approximate time frame for fielding the product.

The high-payoff areas and their road maps are summarized here. For full details, the reader should consult the Biotechnology and Biochemistry Technology Forecast Assessment in the appropriate volume of STAR reports.

- *Deployable bioproduction of military supplies* will use bioproduction methods to produce food, fuel, potable water, explosives, and perhaps ammunition components from indigenous feedstocks. Ultimately, the feedstocks could be as simple as air, water, a carbon source (such as biomass), and sunlight or another common energy source. These theater-based production units would significantly shorten the logistics tail of deployed forces.

The key product identified for this area was deployable bioproduction of fuels. The objective would be a portable unit capable of small-scale (e.g., a few gallons a day) production of an engine fuel. It would be used by small units isolated from regular logistics support, such as Special Forces operations. Near-term Army investments in gene technologies and biomolecular engineering could result by 2002 in laboratory-scale expression of a selected nonpetroleum-based, oxygen-rich fuel. Investment in bioproduction technology beginning in 2005 could result in deployable bioproduction by 2020.

• *Biosensor systems* will include novel ways of coupling electronic or photonic components with biosensors (Figure 3-17). The key products selected for forecasting were a multithreat, deployable detector array, to be fielded after 2020, and integrated bioelectronics. Army investments would be required in the next decade in gene technologies, biomolecular engineering, and biocoupling.

• *Enhanced immunocompetence for personnel* would manipulate the genome of the soldier's white blood cells to confer immunocompetence against diseases and CTBW threats. The response of white cells would be altered to provide not only specific recognition of antigens, as in current vaccination, but also responsiveness to classes of antigens and their potential variants. Troops would be "immunized" in this way just prior to deployment. Gene libraries for immunogens and immunocompetence enhancers would be developed for relevant diseases and known CTBW agents. The capability to immunize troops in this way by 2020 would require immediate Army investment in gene technologies and later (after 2000) investment in targeted delivery systems. Advances in biomolecular engineering would be relevant but would not require Army investment.

• *Novel materials* will result from the use of biomolecular engineering to design new molecules rather than simply borrowing or adapting natural genes to produce naturally occurring molecules. The key products in this area were specialized lubricants, adhesives, and coatings; adaptive camouflage; and multithreat protective clothing against CTBW, electromagnetic radiation, and ballistic impact. The relevant technologies would include gene technologies, biomolecular engineering, bionics, and bioproduction technology. Army investment in biomolecular engineering by 1995 would allow for novel

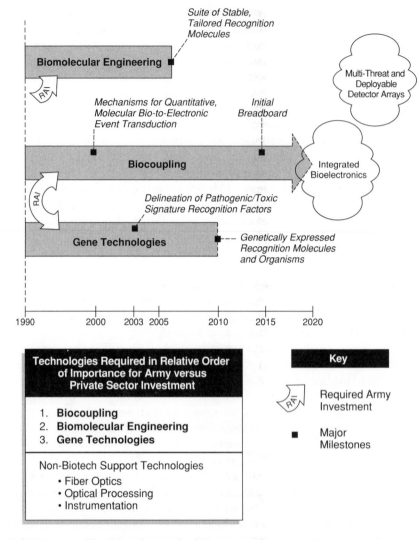

FIGURE 3-17 STAR road map for biosensor system applications.

molecules to be designed by 2003. Immediate Army investment in bionics would enable prototypes of systems and materials based on these novel substances by 2005. Investment in gene technologies after 2000 would enable efficient expression of the novel substances by 2010. The adaptive biocamouflage was projected for fielding around 2020; the other two products would be ready after that.

• *In-field diagnostic and therapeutic systems* will reduce casualties due to disease and CTBW threats. Gene technologies, biomolecular engineering, bioproduction technology, targeted delivery systems, and biocoupling technology will all be required. The key products identified by the Technology Group were rapid, specific diagnosis of symptoms, to be fielded around 2020, and an even longer-term system that would include countermeasure selection and its bioproduction. Immediate Army investments would be needed in gene technologies and biomolecular engineering. Investment would also be needed, beginning in 2005, in targeted delivery systems.

• *Extended human performance* refers to direct coupling of the human central nervous system to machines and other uses of bionics and orthopedics. The required Army investments would be in gene technologies, biocoupling, and bionics. Performance-enhancing compounds and procedures could be fieldable around 2020; bionically linked man-machine systems would become possible around 2030.

• *Antimateriel (soft kill) products* will disable propulsion systems, change the characteristics of soil or vegetation, or degrade warfighting materiel. The Technology Group specified only materiel and terrain as targets for these weapons. Gene technologies, biomolecular engineering, targeted delivery systems, and bioproduction technology would be required. Products to target supplies could be fielded soon after 2010. Antimachine products would come later.

ADVANCED MATERIALS

TFA Scope

The Technology Group on Advanced Materials assessed the following areas of technology:

• *Resin matrix composites* use a polymer-forming organic resin, such as a thermoplastic, in which other structural materials are embedded.

• *Ceramics* technologies produce reaction-formed ceramics, cellular ceramic materials, ductile-phase toughened ceramics, fiber-reinforced ceramics, and diamondlike coatings.

• *Metals* technologies work with special steels (e.g., high-strength, laminated, and steel-based composites); light metal alloys; metal matrix composites; and heavy metals (e.g., tantalum, uranium, and tungsten alloys).

• *Energetic materials* are the basis of explosives, propellants, and pyrotechnics.

Technology Findings

General Findings

This Technology Group expects five materials technologies to have major importance for the Army and recommends them as strategic technologies for special funding consideration: affordable resin matrix composites, reaction-formed structural ceramics, light metal alloys and intermetallics, metal matrix composites, and energetic materials. Reasons for this selection are given under the individual headings below.

The combined benefits of resin matrix composites, advanced ceramics, and light metals will make possible a new breed of ground vehicles. They will be lighter, less costly, and transportable by air, while also being hardened against ballistic attack and more compatible with techniques for low observability.

The Technology Group also identified three pervasive trends in materials science and technology. These trends involve increased use of (1) supercomputers to design materials and to model performance, (2) technology demonstrators to ease the transfer of new materials or processing techniques from the laboratory to the field, and (3) materials and structures designed to serve more than one purpose where multiple layers of single-purpose materials had been used.

The TFA ends with a summary forecast of advances in armor materials available for systems in the near term (up to 15 years) and the far term (30 years and beyond). For this summary, armor applications were divided into three categories: protection of the individual soldier, protection of aircrew and critical aircraft components, and protection of combat vehicles. In the near term, the following advances are expected:

• Ceramic armor will be cost efficient as well as ballistically efficient.

• Composite structural armor will improve in performance, while manufacturing and maintenance costs decrease.

• Advanced aluminum alloys and aluminum-matrix composites will increase performance and reduce armor weight for light-armored vehicles.

• High-performance ceramic glasses will improve transparent armor and dilatant armor for defeat of shaped charges.

• Microstructural texturing will increase the ballistic performance of steel and other metallic armors.

• Polymeric fibers will provide enhanced protection for the individual soldier.

In the far term, the above improvements will be extended and the following new capabilities will emerge:

• Armor materials will be designed for multiple functions (ballistic protection, signature reduction, etc.).
• Biologically engineered fibers will enhance ballistic protection for the individual soldier.
• Nonlinear models for dynamic systems will lead to major advances in the design of materials for ballistic protection.

Resin Matrix Composites

Recent major breakthroughs in processing will significantly reduce the cost of resin-based organic composites. Conventional processing involves the application of heat and pressure, typically through use of an autoclave. Heat transfer depends on relatively slow conduction and convection mechanisms. A nonconventional alternative is electromagnetic curing, which uses microwave-absorbing features of the matrix resin (pendant groups attached to the polymeric backbone) to generate heat evenly and immediately throughout the material. The heating to set the resin can be done in minutes instead of hours. In addition, these advanced processing methods will allow complex parts to be made in one operation. Key Army applications are ground vehicles and in situ structural reinforcements.

Ordered polymers make use of the inherent strength of the carbon-carbon bond by increasing the density of such bonds in the material. If adjacent polymer chains can be more closely aligned, the matrix's mechanical properties can be substantially improved. The chemical backbone of the polymer can also be modified to make the chain itself more rigid. Incorporating these *rigid-rod* polymers within another host matrix gives a molecular-level reinforcement.

Composites that are reinforced in this way, at the molecular level, are easier to process and can be easily fabricated into components with complex geometries. Body armor for the individual soldier is one area of potential applications. Also, because organic polymers are generally transparent to electromagnetic radiation (e.g., from radars), a structure fabricated from polymer-reinforced composites has low-observable characteristics.

"Toughness" refers to the ability of a material to absorb energy with minimal damage and to resist crack propagation. In the development of organic matrix composites, *high toughness* has often conflicted with competing requirements for high strength and stiffness. Anticipated future advances in the molecular engineering of polymer structure and in materials engineering of matrix composition will result in matrix materials that are both tough and strong (Figure 3-18). Thermoplastic matrices are tougher than thermoset matrices. In the composite material, the choice of reinforcement further affects toughness. The gap between the toughness of metals and that of organic matrix composites is decreasing; further research may yield organic composites with the toughness characteristic of metals.

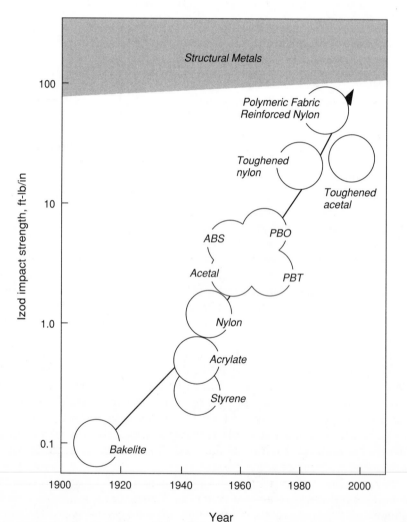

FIGURE 3-18 Toughness levels of organic matrix composites as measured by Izod impact strength.

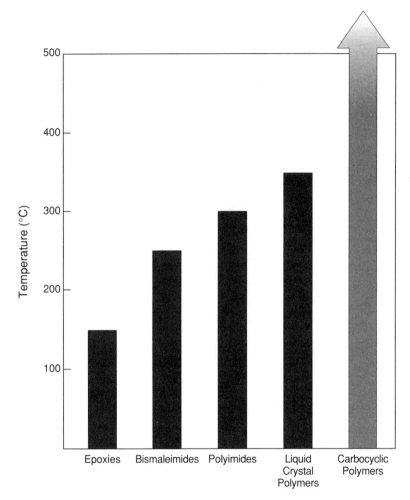

FIGURE 3-19 Continuous-use thermal stability of polymeric matrix materials.

The stability at high temperature of organic composites has also been improving. Molecular engineering of the polymer backbone has raised the continuous-use temperature from 175°C for epoxy to more than 400° for ladder polymers consisting of joined carbocyclic or heterocyclic rings (Figure 3-19). Biosynthesis may provide a cost-efficient production method for the resin monomers with the structure needed to form these polymers. The high-temperature composites made possible by these resins will be important for components of vehicle power plants and weapons. In particular, substituting polymeric materials for metallic components in the hot areas of engines can reduce the heat signature of vehicles.

A *smart composite* has active or passive sensing elements embedded in the matrix. Passive sensor elements, such as optical fibers, can be used to detect temperature, pressure, and stress levels within the composite, either during its formation or when it is in use. During manufacture this information can be fed back to the processing control system. Interrogation of the embedded sensors during the structural lifetime of the composite can detect deterioration and combat damage. Active elements can be used to alter physical and structural properties of the composite during use. For example, vibrational damping characteristics of a composite beam can be varied by embedding a dispersed fluid that changes in viscosity in response to an electric current passing through embedded conducting fibers. Another potential use for active smart composites is in gas turbine engines, where properties of the compressor section could be modified to correct for wear, damage, or mission requirements.

Increased use of organic composites raises the question of bonding structural components of either the same or different composition. *Adhesive bonding* has major advantages over mechanical fastening when the bond is intended to endure for the life of the structure. Knowledge of the mechanisms for cohesive failure within the bonding layer is adequate to predict failure rate, but failure at the interface between layers is less understood. Further research in this area would benefit applications in battlefield repair of structures and reduce the time and cost of logistical support for maintenance and repair.

Ceramics

In *reaction-formed ceramics*, reactions among constituent materials during consolidation results in final structures with the same shape and dimension of the preconsolidation mix. So the process step that creates the ceramic material also shapes it to its near-final form. In addition, fiber-reinforced ceramics are easier to produce, because shrinkage of the matrix away from the reinforcing material is eliminated. This ability to preform ceramics and ceramic-composite articles will greatly reduce costs for ceramic applications in armor, antiarmor, and diesel engines.

For the near term (to 15 years out), the most important issue for reaction-formed ceramics will be control of the reaction rate. Controlling this rate requires continued research on reaction mechanisms and accumulation of experiential data. The intrinsic reaction time of mechanisms now in use ranges from slow reactions that take days (e.g., chemical vapor infiltration) to near-explosive reactions performed in seconds (e.g., self-propagating high-temperature synthesis, or SHS). The techniques in the intermediate range, which take minutes to hours, appear most promising for the manufacture of complex-shaped components.

The Technology Group forecasts that in the far term (30 years out), reaction-formed techniques will begin replacing conventional sintering technology even for well-established, low-cost items. This technology will affect military applications such as armor, power and propulsion, gun barrels, missile guidance, and the packaging of electronic systems.

Cellular ceramics are porous rather than monolithic; they have a foamlike structure. In assessing the properties of a cellular ceramic, the appropriate comparison should be with materials it might replace rather than with the pure, dense form of the same ceramic. For example, a ceramic foam may have superior characteristics to a structural organic polymer, including greater tolerance for high temperature and chemical exposure. In some cases cellular ceramics may even have greater specific strength than a pure monolith of the same material.

For the near term, the Technology Group forecasts improved understanding of how to design a cellular ceramic with optimum properties, a growing body of data on these materials, development of processes for engineering the cellular microstructure, new coatings for cellular ceramics, and initial use of cellular ceramics in structural and electronic applications. For the far term, from 30 years out, the Group foresees cellular ceramics replacing dense structural ceramics and metals for both land-based and space applications. Their low dielectric constant will make them well suited for electronics packaging. Military uses may include heat shields in engines, high-temperature structural materials, thermal management in high-power electronics systems, and as a technique for low observability.

Ceramics can be toughened by the inclusion of a *dispersed ductile phase*, such as a metal or metal alloy. New processing methods, which make manufacturing easier, can be applied to a wide range of ceramic-metal combinations. While these "cermets" are naturally well suited for low-temperature applications, further work is needed on high-temperature composites. For the far term, the Technology Group expects cermets to be widely used in large-scale structural applications. Military applications, which will make use of the low manufacturing cost, light weight, scale-up potential, and unique properties of cermets, will be in such areas as armor, gun tubes, structures for high-power electronics, and moderate-temperature engine components.

In *fiber-reinforced ceramics*, either fibers or whiskers of a ceramic (silicon carbide) or carbon are embedded in a ceramic matrix to improve the strength, fracture toughness, modulus, or thermomechanical properties of the matrix. Also, the composite often shows "graceful," or gradual, fracture rather than catastrophic failure. Either reaction-forming (see above) or pressure methods of manufacture are necessary to prevent shrinkage of the matrix away from the reinforcing fibers. Another problem is susceptibility to oxidation at high temperature.

For the near term, the Technology Group expects fiber-reinforced glass and glass-ceramic composites to be used for relatively small components. Developments will focus on improving the fiber materials and using fiber coatings to overcome high-temperature oxidation and tailor interfacial mechanical properties. In the far term, larger structures will become feasible. Near-net shape-processing methods (such as reaction forming) will predominate, multidirectional fiber weaves will be available, and joining techniques will mature. Fiber-reinforced ceramics will find military applications in armor, propulsion and power, metal-cutting and metal-forming tools, gun barrel technology, and spacecraft structures.

Diamond and diamondlike coatings are thin films deposited on a bulk substrate. The coatings give hardness, high thermal conductivity, infrared transparency, and other properties associated with diamond. Among the many potential military applications are abrasion-resistant coatings for sensor windows and other optics, high-power transistors and optically activated switches for high-power microwave or millimeter-wave sources, wear-resistant coatings for bearings and journals, substrates and insulating films for high-power electronics, and inert wear-resistant coatings for medical implants.

Metal Technologies

Even though *ferrous metal* technology is relatively mature, significant advances in improving properties are possible. Research into the relations between structure and particular properties, such as tensile strength and toughness, can open the way to process changes. A recent example is research into the role of sulfide inclusions in toughness; advanced techniques for producing low-sulfur steels have been able to increase toughness at a given strength. The Technology Group foresees further developments, following this pattern, with respect to mechanisms of failure ahead of a crack tip, further work on toughness, decreased susceptibility to hydrogen embrittlement in gear and bearing steels, and surface treatments to improve resistance to wear and corrosion.

Laminated steels offer the possibility of combining the high wear resistance and strength of ultra-high-carbon steels with the high impact-resistance (toughness) of lower-carbon steel or other materials. The processing characteristics of such laminates also appear favorable.

Recently, the slow, evolutionary rate of advance in *light metal alloys* has been revolutionized by the development of an aluminum-lithium alloy with greater specific strength and stiffness than steel. Processing innovations have also raised the commercial potential of improved and new light-metal alloys. For instance, powder metallurgy for rapidly solidified (PM/RS) aluminum alloys offers potential for materials that will compete with titanium alloys for high elastic modulus and strength at high temperature. Although problems in areas such as fatigue and fracture toughness still need resolution, these alloys are likely to find important applications in engines and robot vehicles.

PM/RS has also increased the prospects for magnesium-based alloys. The improvements in alloy properties are similar to those of PM/RS aluminum alloys. Applications for structural members in helicopters and other lightweight vehicles appear to be their major uses. Similarly, new lighter *intermetallics* will eventually replace heavier nickel or cobalt-based alloys.

Advances in knowledge of composites based on steel and aluminum matrices will benefit the entire area of *metal matrix composites*. The Technology Group forecasts significant expansion of this area in the next 10 to 20 years, with revolutionary effects on the properties of steels, titanium, magnesium, intermetallics, copper, and heavy metals, as well as aluminum.

In addition to the advances in ferrous metal techniques described above, properties of steel such as wear resistance and tolerance of high temperatures may improve through the use of *steel composites*. Particles of a wear-resistant compound, such as TiC or TiN, are added to the steel. Research in this area is needed, as the effects of such particulates on high-temperature properties are not known. Stainless steels, which resist oxidation at high temperature, may be significantly strengthened by addition of these high-strength particulates. Applications with military significance include gas turbine engine components, transmission housings, and advanced gun systems.

Similarly, *aluminum-based composites* have ceramic particulates or whiskers added to the aluminum matrix. Addition of these reinforcements significantly increases stiffness while reducing the coefficients of thermal expansion and thermal conductivity. Composites with the stiffness of titanium are projected. Here, too, further research will be needed on the effects of composites on other properties of the matrix metal. Other reinforcing materials and processing techniques also should be investigated. Some recent work indicates significant improvement of toughness in aluminum matrix composites by controlling the microstructure, but further investigation is needed, as is research on strain rate.

Particulate ceramics also offer great potential for reinforcing other metal and intermetallic matrices. The Technology Group cites one new process that appears usable for the addition of carbide, nitride, or boride ceramic particles into aluminum, copper, titanium aluminide, or nickel aluminide matrices. The resulting composites have specific properties superior to the superalloys, based on nickel or iron, now in use; they also offer a 50 percent weight reduction. Another advantage of metal matrix composites is their potential for advanced forming techniques, such as superplastic forming. The Technology Group forecasts applications of these composites in lightweight armor, missile components, rotating structures, gun barrels, and electrically powered guns.

The Technology Group focused on two areas of *heavy metals* technology: tantalum warheads for shaped-charge penetrators and alloys of depleted uranium or tungsten for use in kinetic energy projectiles. Materials research on tantalum penetrators focuses on metallurgy to produce penetrators that respond well to the explosive deformation processes that form them and do not break apart before reaching the target. Code-research groups are working on mathematical models to bridge the gap between the microscopic results of metallurgy and macroscopic experiments in shock-wave physics.

Most of the current research on depleted uranium alloys aims at improving the methods of thermal and mechanical processing of this highly anisotropic metal. Two promising areas are new alloy casting techniques and the use of PM/RS technology (described above for light metal alloys). Tungsten alloys continue to underperform depleted uranium alloys as kinetic energy projectiles against heavy armor. However, tungsten alloys work as well as depleted uranium against light armor; further research may improve their performance against heavy armor to the point where they are fully acceptable as an alternative to depleted uranium.

Energetic Materials

The Technology Group identifies two emerging technologies that are likely to change Army energetic materials significantly. New organic cage compounds will have higher material densities, and there-

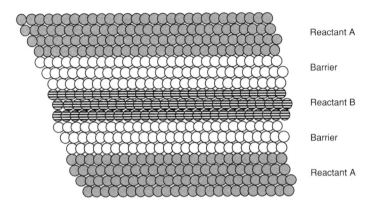

Reactant A

Barrier

Reactant B

Barrier

Reactant A

FIGURE 3-20 Thin layers of energetic reactants A and B could be separated by inert barrier layers with micron-range thickness.

fore more explosive power, than conventional organic explosives such as HMX and RDX (cyclotetramethylene tetranitramine and cyclotrimethylene trinitramine, respectively). Entirely new energetic materials are also possible by separating thin layers of inorganic reactants with inert barrier layers (Figure 3-20). The layer thicknesses would be on a micron scale.

Research also continues in methods to make energetic materials less sensitive to uncontrolled stimuli (fire, ballistic impact, or explosion of nearby munitions) without sacrificing their performance. The Technology Group forecasts progress in understanding the phenomena of sensitivity during the next decade, including computer modeling capability to design insensitive materials with high energy density before synthesizing them in the laboratory. Within 20 years, processing technology will allow smaller particle size and better control of the interaction between energetic material and the matrix of binder and plasticizers.

In the processing of energetic materials, biotechnology is likely to become important not only for the biodegradation of hazardous waste products but also for the synthesis of energetic molecules.

PROPULSION AND POWER

TFA Scope

The Technology Group on Propulsion and Power has assessed the following areas of technology:

• *High-power directed energy* covers the technologies for high-energy lasers (beamed radiation at optical frequencies) and directed beams of radio frequency, microwave, and millimeter-wave radiation.

• *Propulsion technologies.* The Technology Group divides propulsion technologies into four categories: missile propulsion, air vehicle propulsion, surface mobility propulsion, and gun or tube projectile propulsion.

• *Battle zone electric power* covers generators of continuous electric power, generators for pulsed and short-duration power, and energy storage and recovery for ultimate use as electric power.

Technology Findings

High-Power Directed Energy

High-power lasers, microwaves, and directed energy weapons offer the opportunity to disable or destroy enemy systems. Four promising laser types currently under investigation for this purpose are chemical lasers, free-electron lasers, ionic solid state lasers, and coherent diode laser arrays. Potential laser applications include ground-based ballistic missile and antisatellite defense, air defense, antisensor, and antipersonnel use. High-power microwave (HPM) technology offers the opportunity to physically damage enemy systems in the same types of applications as optical-wavelength lasers. As an energy source, beamed microwave power could also be used to keep remotely piloted aircraft aloft indefinitely. In the findings below, these various directed energy beam technologies will be referred to generically as *high-power directed energy* (HDE).

The Technology Group selected five HDE technologies as high-leverage areas for Army support: (1) ionic solid state laser arrays; (2) coherent diode laser arrays; (3) phase conjugation for high-energy lasers; (4) millimeter-wave generators (high-power); and (5) multiple-beam, klystron HPM technology. Each received a detailed assessment and a figure-of-merit scoring.

In considering high-energy lasers, the Technology Group focused on requirements in four areas of application:

• For *ground-based ballistic missile defense*, the laser source produces a beam that is reflected by space-based relay mirrors onto the target missile while it is still in its boost phase. Lasers for this application must have very high power to deliver sufficient kill power in a short time over a long range. Wavelength selection is important to minimize power loss to the atmosphere.

- *Antisatellite* laser requirements are similar to those for ballistic missile defense. For targets in low-to-medium orbit, the power required may be an order of magnitude lower, because the targets are softer and distances are shorter.
- *Air defense* against aircraft and cruise-type missiles requires a laser that is effective at less than 10 km. The power requirement is an order of magnitude less than for antisatellite weapons. To be practical, the laser system must be transportable, perhaps even highly mobile. It must also be able to fire repeatedly at multiple incoming targets (100 to 200 shots at 2 to 3 s each) at multiple incoming targets before fuel reloading is needed. (In the terminology of conventional ballistic weapons, it must have a large magazine capacity.)
- *Antisensor* lasers must be capable of crazing or destroying the optics, detectors, or other elements of sensor systems. The power requirement is moderate. The laser wavelength may be either in or out of the operating bandwidth of the threat sensor, but the source, together with its beam control and fire control subsystems, must be in a package suitable for mobile operation at or near the forward line. The requirements for *antipersonnel* systems are the same.

Ionic solid state laser arrays were selected as a high-leverage technology for the Army. Their output takes the form of pulses with high peak power. They can be built for either low pulse rate and higher peak energy (glass host) or high pulse rate and lower peak energy (crystalline host). At present, they have efficiencies of 2 to 5 percent when flashlamp pumped and 10 to 15 percent when pumped by (incoherent) diode laser arrays. They are currently used for range finding, target designation, remote sensing, and communications.

For weapon applications, ionic solid state lasers must be scaled well beyond their current performance levels. The Technology Group forecasts improvements to possibly 30 percent efficiency and drastic cost reduction through pumping with diode laser arrays. Complete tunability from the mid-infrared to the ultraviolet region is also possible, although R&D work is required. Also, to make ionic solid state lasers acceptable for tactical applications, the cost of the diode laser arrays used to pump them must be reduced by one or two orders of magnitude. The Technology Group forecasts that crystalline-host lasers could achieve average powers greater than 500 W per aperture, delivering 1 to 4 J of energy at pulse rates of 200 to 500 Hz. Glass-host lasers could achieve outputs of more than 1 kJ at repetition rates of 1 to 3 Hz. Coherent coupling could raise outputs to levels 10 to 30 times higher than that.

Coherent diode laser arrays, another high-leverage technology choice, are tunable by adjusting the composition and temperature of the semiconductor whose bandgap transition is the radiation source. The Technology Group forecasts that power levels of 10 to 1,000 W per modular unit, with energy fluxes as high as 1 kW/cm^2, can be achieved in the next 5 to 10 years. Efficiencies of 50 percent appear achievable. The main development issues for extending this technology to weapon system power levels of tens to hundreds of kilowatts are (1) extending the mechanisms for phase locking hundreds or thousands of individual diode laser modules into an extended coherent array and (2) managing the waste heat generated by high-power operation.

The Air Force is currently conducting research on coherent diode laser arrays. The Technology Group recommends that the Army not only monitor the Air Force's projects but also pursue complementary work.

In addition to these two high-leverage laser technologies, the Technology Group reviewed two other laser source technologies: *chemical lasers* and *free-electron lasers* (FELs).

The most highly developed chemical laser is hydrogen fluoride/ deuterium fluoride (HF/DF). Its major advantage is that electric power is not required for lasing, which is produced from direct chemical reaction. However, to be operated in continuous mode, the laser cavity must be at low pressure; in pulsed mode it may be operable at atmospheric pressure. A major development forecast by the Technology Group is short-wavelength (visible or ultraviolet) chemical lasers.

The mode of operation of FELs, as well as ongoing Army research on them, has been discussed above for the TFA on Optics, Photonics, and Directed Energy. The main future use of FELs will be in ground-based ballistic missile defense and antisatellite weapons. This Technology Group does not consider them practical for mobile hard-kill weapons, although they may prove usable for antisensor countermeasures and soft-kill weapons.

Among the laser beam control technologies reviewed by the Technology Group, *phase conjugation* was selected as a high-leverage technology for the Army. It is a relatively new, nonlinear optical process that can be used to correct dynamic beam aberrations in real time. It can be used for beam cleanup, beam combination, and array phasing. It is simpler to integrate into existing systems and has a faster time response than conventional adaptive optics. The capabilities of phase conjugation have been demonstrated on pulsed and continuous-wave lasers using a variety of source methods. There are two processes in use, one suitable for high-power lasers, the other for beams with low or moderate power. By relaxing optical tolerances and replacing conventional adaptive optics, phase conjugation can reduce both the cost and complexity of moderate-power and high-power lasers.

As a method for combining beams, phase conjugation can conceivably allow large increases in laser output power. An n-by-n array of semiconductor diode lasers, each of power P, can in principle be phase-conjugated to produce an output beam with intensity n^2P. Phase conjugation may also be applicable to optical countermeasures; a threat beam can be phase conjugated, amplified, and returned against its source, causing the source's destruction. By 2020, phase conjugation cells to track targets automatically and point directed energy beams could be well developed.

Radio frequency, HPM, and millimeter-wave weapons. The advanced electronics technology of the modern battlefield—advanced sensors, smart weapons, and autonomous systems—typically use radio frequency, microwave, or millimeter-wave signals ranging from 100 MHz to 300 GHz. The primary tactical significance of high-power electromagnetic pulse energy beams that operate in this same region lies in their potential to damage the electronics necessary for hostile sensors and systems to function.

Of eight technology options in this frequency regime that the Technology Group reviewed, it selected two as high-leverage technologies for the Army: coherent multiple-beam HPM energetic pulse klystrons and millimeter-wave sweeping-frequency generators.

An *HPM pulse* kills its target by coupling to electronic circuitry that it accesses through either functional openings or cracks in the target's shielding. The thermal energy from this coupling burns out device junctions. Although this nonselective mechanism does not require tuning to the threat system's operating frequencies, it does require a relatively high power density at the target. Together with the desired long range of the HPM weapon, this implies a high peak-power requirement for the beam source.

The klystron tube is basically an amplifier technology; a microwave-modulated beam of high-current electrons generates output microwave power with as much as a 50-dB gain over the input power used to modulate the beam initially. A device with a single electron beam at 500 kV and 300 A can produce 60-MW peak microwave power in an output beam that pulses at more than 100 Hz. Current research is focusing on the use of multiple-beam klystrons. This research is being aided by simulation codes, which allow comparison of design alternatives prior to selecting the best design for demonstration.

Assuming that adequate funding and technical manpower are avail-

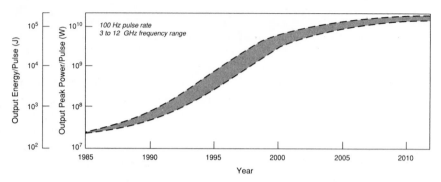

FIGURE 3-21 Technology forecast for a multiple-beam Klystron high-power microwave system.

able, the Technology Group forecasts that a multiple-beam klystron HPM weapon with peak power of several gigawatts can be fielded by 2000 (Figure 3-21). The peak power could be in tens of gigawatts by 2010. Achieving these goals will require substantial development beyond the current state of the art for components such as power conditioning systems, modulators, and thermionic electron cathode sources, as well as the klystrons. The forecast includes an outline and funding levels for the development effort needed to achieve fielded systems by the projected dates.

A *millimeter-wave sweeping-frequency generator* would produce tunable-frequency pulses of radiation in the range of 15 to 300 GHz. The kill mechanism is to engage target apertures and sensors to cause electromagnetic damage. Although the required energy delivered on target can be much less than in the HPM mechanism, the beam frequency does need to move through the operating range of the aperture or sensor. Multiple gyrotron oscillators can be used as beam sources to cover the range from 15 to 50 GHz, while FEL lasers can be used in the range of 50 to 300 GHz. With a high-power narrow beam, the effective range of this weapon can be hundreds of kilometers. Other requirements of an effective system are a heavy-duty cycle, at least 1-MW power to defeat smart weapons with millimeter-wave imaging or infrared guidance, and a frequency sweep range of 30 to 300 GHz.

The Technology Group forecasts that gyrotron technology will produce megawatt-level continuous-wave power by about 1995, but it will be at fixed frequencies (Figure 3-22). In the same time frame, FEL technology will provide 1-MW power levels with limited tuning capability and pulse rate. The FEL technology for tunable frequency

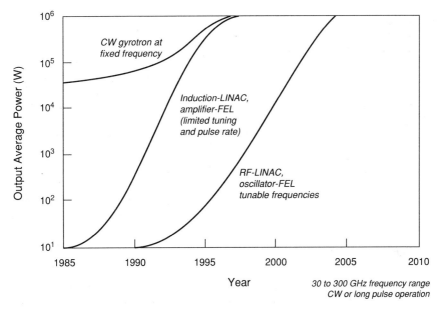

FIGURE 3-22 Technology projection for millimeter-wave FEL sweeping-frequency generators.

sweep across the entire 30 to 300 GHz range with continuous-wave 1-MW power can be achieved by 2005. These projections assume a program funded at the levels recommended by the Technology Group.

Propulsion Technologies

The Technology Group assessed multiple technology options in each of the four propulsion categories, as shown in Table 3-2. It recommends four options, one from each category, as high-leverage technologies for special consideration by the Army. These options are shown in bold and underlined in the table.

The Army's future *missile propulsion* technology must be adequate for new generations of smart-to-brilliant missiles. To achieve the range variation, minimum flyout times, targeting flexibility, and accuracy required for both offensive and defensive battle zone missions, these missiles must be capable of extremes of energy management and maneuverability. Therefore, the propulsion technology must provide the missile designer with broad options for peak thrust levels, thrust-time profiling, and multiple restart with pulse durations as short as tens of milliseconds.

TABLE 3-2 Propulsion Technology Options

Missile Propulsion	Surface Mobility Propulsion	Air Vehicle Propulsion	Gun or Tube Propulsion
Solid rockets	Diesel engine	Gas turbine jet	Solid charge
Liquid rockets	Gas turbine engine	Propeller-driven	Traveling charges
Gel-propellant rockets	Transmission technology	Ram/scram jet	Liquid-propelled
		Solar-powered	*Electrothermal*
Hybrid rockets	*Electric drives*		
		High-power microwaves	*Electromagnetic*
Turbine engines	Integrated systems		Hydrogen cannon
Ram/scram jets	Hybrid-electric		
			Ramjet-distributed
Ducted/air-augmented	Ground effect		

Of the rocket propulsion technologies assessed by the Technology Group, *gel propellants* offered the most potential for new technology. They are generally less sensitive and safer to handle and store than either liquid or solid propellants. Under high pressure, they shear like liquids, so they have the superior energy management characteristics of liquid propellants (multiple stops and starts, proportional throttling). Their performance characteristics may be even higher than liquids, which in turn are generally higher than solids. Gels also aid in signature management, because chlorine and carbon products can be eliminated from the motor exhaust.

Solid propellants will see evolutionary improvement in specific impulse and other performance parameters, as new systems such as glycidal azides or high-strained carbon bond fuels substitute for the current nitrate-ester-plasticized polyethane (NEPE). However, energy management will remain difficult to implement with solid propellants.

Among the *air-breathing missile propulsion* technologies that the Technology Group assessed, turbine engines and ducted or air-augmented rockets showed the most promise for significance to the Army in 2020. The Technology Group expects that progress in these technologies will follow as a matter of course from various programs already in place among the military services. Substantial decreases in specific fuel consumption and increases in thrust per unit airflow are expected for future gas turbines.

For *aircraft propulsion*, foreseeable developments in gas turbines offer the potential to double current performance by increasing the power-to-weight ratio and thrust-to-airflow ratio and by reducing fuel consumption. Advanced materials will reduce the weight of structures and components. (See Advanced Materials TFA for details.) Current programs will improve the performance of the Army's rotary wing and fixed wing manned aircraft. The Technology Group recommends that the Army rely largely on the ongoing Integrated High-Performance Turbine Engine Technology (IHPTET) program.

The high-leverage air propulsion technology selected by the Technology Group is the use of *HPM (high-power microwave) to power high-altitude surveillance UAVs* from ground transmitter sources (Figure 3-23). The components most in need of Army-specific development are the "rectenna," which is located on the underside of the vehicle's wing to capture microwave radiation and convert it to electric current, and the ground-based transmitter. To lessen the UAV's signature when it is in hostile airspace, it can carry rechargeable batteries and use supplementary photovoltaic cells on the wings' up-

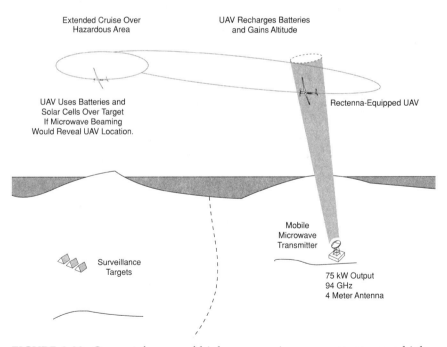

FIGURE 3-23 Concept for use of high-power microwaves to power a high-altitude surveillance UAV.

per surface. Periodically, it would fly back to safer airspace to have its batteries recharged from an HPM transmitter station.[4]

In the propulsion category of *surface mobility*, the Technology Group assessed primary power production (engines), methods of power transmission or distribution, and mechanical subsystems and components. It also assessed three general conceptual approaches to surface vehicle propulsion. Two of these general concepts, the Integrated Propulsion System (IPS) and hybrid-electric propulsion systems, received special consideration as aspects of the Group's overall high-leverage technology for surface mobility: a system designed under an IPS approach, having an advanced diesel or gas turbine engine and either all-electric or hybrid-electric power distribution.

The third general mobility concept, *ground-effect machines*, was not evaluated in detail by this Technology Group. (The technology is discussed at some length in the Mobility Systems Panel report.)

The likelihood that Army combat vehicles will in the future face enemies that have numerical superiority places a premium on vehicle survivability. One response is to improve armor; another is to produce more mobile, agile vehicles that are lighter in weight, have smaller silhouettes, and incorporate low-observable technology. The technology to enable the second response depends on lightweight low-volume propulsion systems (Figure 3-24).

For *primary power*, the Technology Group forecasts evolutionary progress in diesel engine performance through turbocompounding and stratified-charge combustion. In turbocompounding the engine exhaust gases are used to drive an input-air compressor (turbocharging) that is also linked to the flywheel. Stratified-charge combustion uses a spark-ignited precombustion chamber to ignite a leaner fuel-air mixture in the main chamber. The other promising option for future engines is an advanced gas turbine, pursuing the evolutionary line represented in the M-1 class Abrams tank engine.

The Technology Group foresees *electric drive* as the most promising power distribution system for the far term (circa 2020), despite negative assessments of it by the Army in previous decades. The Group's support for either an all-electric or hybrid-electric system takes into account dramatic improvements in all areas of electric-drive technology during the 10 years since the last Army review. The advantages

---

[4]This propulsion concept was also assessed by the STAR Airborne Systems Panel, which judged it to be poorly suited for Army applications. The strong, continuous RF signal could be easily detected and the transmitter attacked. The aircraft's flight pattern would also be restricted to keep it close to the broadcast array.

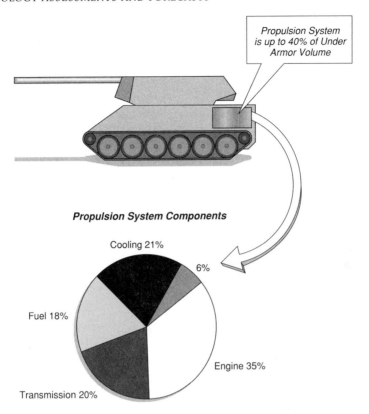

FIGURE 3-24 Propulsion system size in typical fielded vehicles.

relative to other power transmission approaches include (1) improved weight distribution, since components are modular; (2) individually driven wheels or track drive sprockets, eliminating complex transmission/differential drive trains; and (3) a common power distribution system for vehicle drive, electrically powered weapon systems (such as an electrically energized hypervelocity gun), and power storage.

A *hybrid-electric propulsion system* includes a mechanical transmission link between the engine and the wheels or track drive sprockets, in addition to the generator and traction motor subsystem for the electric drive.

IPS (integrated propulsion system) is a conceptual approach for applying integrated systems design to propulsion systems, rather than a propulsion method (Figure 3-25). Under IPS all aspects of the

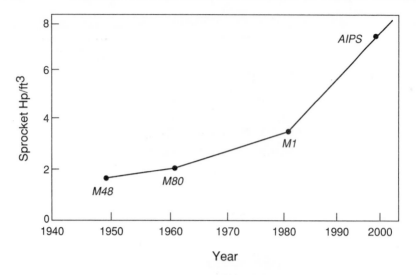

FIGURE 3-25   Projected improvements in power density due to IPS design concepts.

system—engine, transmission, cooling system, air filtration, auxiliary power, inlet and exhaust ducts, diagnostics, signature reduction, and even maintenance—are designed interactively, rather than by small teams designing components in isolation. Currently, IPS is being applied by the Tank-Automotive Command for the design of a diesel-based system and a gas turbine system. These designs should increase power density by at least 50 percent, with a similar increase in horsepower per unit weight.

The Technology Group recommends that the IPS approach also be applied to long-term consideration of a propulsion alternative that incorporates electric drive. For example, a hybrid electric-drive propulsion system offers major gains in total battle zone effectiveness, gains that will be enhanced by electrical energy storage systems with far higher electrical power density than current technology. Two such systems that offer particular promise are advanced batteries (perhaps a fivefold improvement over lead-acid batteries) and flywheels (fourfold improvement or more).

For *propulsion of projectiles from guns or tubes,* the Technology Group reviewed chemically energized propulsion, electrically energized propulsion, and hydrogen cannon (hot gas propulsion). The chemical propulsion options reviewed were modular charge, rocket-assisted projectiles, traveling charge, liquid propellants, and ram guns. The Group selected chemical propulsion by liquid propellants and electrically energized guns as the technologies having the most potential to augment gun capability for the future Army.

Liquid propellants are an evolutionary approach to gun propulsion technology. In the ongoing Army program, state-of-the-art gun barrel and recoil mechanisms can be used; only the breech must be redesigned. Liquid guns with 25-, 30-, and 105-mm rounds have been successfully demonstrated. Liquid propellants are relatively insensitive to shock, are stable in long-term storage, and offer a number of performance benefits. Further evolutionary improvements to the current technology are possible.

Two new types of guns will use electrical energy, in whole or in part. In the *electrothermal chemical (ETC) gun*, as in a gun using conventional propellants, the projectile is accelerated by the high-pressure gases created in the gun tube. In an ETC gun these gases are created by the combustion of normally inert materials at the very high temperature of a plasma, which is created by a pulse of electrical power supplied to the gun breech. Provided the pressures can be tolerated, the ETC gun can achieve projectile velocities that are perhaps 30 to 40 percent higher than can be efficiently achieved with conventional propellants. This is certainly as high as is needed for field artillery systems.

In the *electromagnetic (EM) gun*, the projectile is accelerated by electromagnetic pressure instead of gas pressure. EM guns have been demonstrated with very small projectiles at velocities of 7 km/s. With projectiles of the mass needed for air defense and antiarmor roles, velocities well over 2 km/s have been achieved. Currently, the pulse power conditioning unit for the EM gun is too bulky for compact ground vehicles, but it is expected to be reduced substantially in the next few years.

On current evidence, both ETC and EM guns may have a place in the set of future Army armaments.

Battle Zone Electric Power

The future Army will require electrical power in the battle zone at levels from tens of watts for surveillance and communication to hundreds of megawatts for directed energy weapons. Mobility will be essential. For mobile *continuous-power generation*, the key to substantial weight reduction is to increase the generating and distribution frequency from the current standard of 60 Hz to 400 Hz or higher.

Internal combustion turboshaft (gas turbine) engines, which are already in use for mobile electric power, offer more potential for the future than the alternatives (internal or external combustion piston engines and fuel cells) for mobile continuous-power generation. A turbine running at 24,000 rpm can drive a 400-Hz alternator directly, without the heavy gearing now needed to drive 60-Hz alternators. Continued Army support for the IHPTET program (recommended above for aircraft propulsion technology) can realize the potential of this technology, when coupled with an aggressive effort to advance the technology for high-frequency, lightweight alternators, power conditioners, and distribution system. This effort could raise the power-to-weight ratio for mobile electric power units from the range of 0.05 kWe/kg (kilowatts electric per kilogram) for current 60-Hz gas turbine units to more than 3 kWe/kg.

In power generation and distribution, the use of high voltages can also decrease weight; the conductor weight required for a given wattage decreases as the square of the voltage. Improvements in high-voltage semiconductor devices would allow an increase from the current limit of about 1 kV to levels at which a power-to-weight ratio of 5 kWe/kg would be possible for a mobile electric unit.

As a primary power source, fuel cells would become practical only if a breakthrough occurs that would allow liquid hydrocarbons and air to fuel them.

Directed energy devices and other electrically energized high-power systems of the future Army will require *generators for pulsed and short-duration power* whose average power for the duration of output ranges to hundreds of megawatts. In both the mass and bulk (volume), generators in this class are half prime power unit and half power conditioning unit. For the prime power unit, the technologies with the most promise for the Army were judged to be gas turbine engines (for energy production) and flywheels (for energy storage). For power conditioning, new, molecularly tailored solid state devices and improved methods of heat removal should make possible an order-of-magnitude reduction in weight.

For *power conditioning* in pulsed or short-term generators, the Technology Group sees the development of high-temperature, high-power electronics as a crucial area. In particular, continued evolution along present lines must be pursued for capacitors, inverters, switches, and transformers. For each of these component types, the combination of high voltage, high frequency, and high power requires technology that is beyond the current state of the art but not out of reach.

Energy Storage and Recovery

Reducing the observable signature of power generation units in the battle zone will become increasingly important. Technologies that allow storage of power in low-signature devices, such as secondary batteries or flywheels, will become critical for the short but intense conflicts on future battlefields. For example, mobile systems may move into position using internal combustion engines for locomotion; under battle conditions they would switch to their onboard energy storage devices. Of the storage device technologies reviewed by the Technology Group, rechargeable batteries and mobile (vehicular) flywheels were selected for their broad applicability to Army needs in 2020.

Rechargeable (secondary) batteries capable of a large number of discharge/recharge cycles probably will play a far greater role in the future Army than they have in the past. The current state-of-the-art lead-acid battery needs to be replaced with an innovative technology. The Technology Group forecasts an increase in energy density by a factor of four or five and of power density by two or three for a new battery technology relative to current lead-acid batteries. It projected future (2020 time frame) performance parameters for five battery technologies now in the research stage.

The anticipated advances in *flywheel* technology will come primarily from new composite materials with high ratios of tensile strength to weight (Figure 3-26). These materials will increase energy density

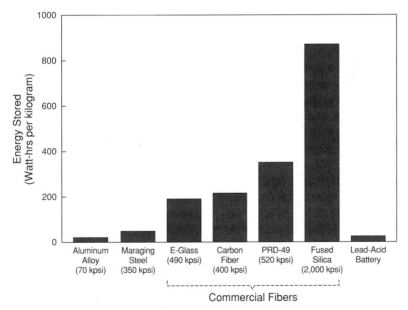

FIGURE 3-26 Storage capacity of flywheels of different composition (tensile strength in parentheses) compared with lead-acid storage batteries. (SOURCE: Richard F. Post and Stephen F. Post, 1973, Flywheels. Sci. Am. 229:17-23.)

by an order of magnitude over flywheels made with high-strength steels. The cost of fabrication for composite flywheels should also decline dramatically over the next 30 years.[5]

ADVANCED MANUFACTURING

TFA Scope

The Technology Group for Advanced Manufacturing focused on the systems aspects of manufacturing rather than individual process technologies. Specifically, the Group reviewed the following topics:

• *Key technologies* include intelligent processing equipment, microfabrication and nanofabrication, flexible computer-integrated manufacturing, and systems management.
• *Applications important to the Army* include distributed and forward production facilities, rapid reaction to operational requirements, and parts copying.
• *Issues in manufacturing technology* include sources of supply, availability of materials and components, military versus civilian R&D industrial preparedness, capital investment and facilities, flexible production schedules, design for manufacturability, and environmental and legal issues.

Technology Findings

General Findings

The focus for technological advances in the next generation of manufacturing will be on the inclusion of information systems with the energy systems and material management systems introduced by previous generations of technology. Instead of having only persons and machines involved in a manufacturing process, automated machinery now includes some form of computer control based on feedback from sensors.

Adding information systems to manufacturing results in major improvements in accuracy, reliability, and quality. For example, auto-

---

[5]The Mobility Systems Panel expressed reservations concerning the prospects for flywheel technology as forecast by the Propulsion and Power Technology Group. Technical difficulties may arise with the energy input and output mechanisms rather than with the flywheel itself.

mated machinery can (1) reduce the cost for increased functionality; (2) enable civilian specifications and quality standards high enough that separate military specifications are not necessary (which implies that civilian production facilities can be used for military production); and (3) significantly decrease the time from concept to deployed system.

Because production lines can be tailored quickly, order quantities no longer must be large to be economical. Variations in products will neither add cost nor reduce reliability. Therefore, it becomes possible to customize weapons and support gear to fit a specific intended use in a specific environment, rather than requiring an item to fit a broad category of conditions for a design lifetime of 10 to 15 years.

By changing response times for resupply from years to weeks, with the ability to customize for current needs, the inventory that must be maintained is sharply reduced.

Technological advances in material transformation processes combine new scientific understanding of the underlying transformations with automated control systems to monitor and control the process. These changes in processing technology will accelerate three trends: (1) the ability to specify the attributes of a material ("designer materials") will broaden to include the ability to design and fabricate "designer parts"; (2) the information subsystems component of larger systems will increase; and (3) the reproducibility of processes and control information will increase the ability to model variations in process variables and predict system performance.

Key Technologies

Intelligent processing equipment can sense (i.e., monitor with appropriate sensors) important properties of the material that are altered by the process, and it has the intelligence to control changes in these properties. Although industrial robots are the most visible component of this technology, to perform they must be coupled with sensor systems and intelligent control systems.

Microfabrication and nanofabrication involve manipulating and fabricating materials at the microscopic or atomic level, respectively. The next generation of integrated circuit chips will require these techniques (see TFA on Electronics and Sensors, above, and Basic Sciences, below). Microscopically applied films and surface treatments are used not only in microelectronics fabrication but also in metallurgy for low-friction bearings and other special characteristics. The potential for low cost and high sensitivity in new devices with microscopic dimensions will make possible microsensors for measuring flow, pressure, chemical concentrations, and other parameters in mechanical, medical, and environmental applications.

Flexible computer-integrated manufacturing (CIM) applies information systems technology to the levels of manufacturing integration above the level of intelligent processing equipment, which applies information control to a single process or workstation. A group of workstations, constituting a factory cell, can be organized around a set of related tasks or functions. Cells are combined into factory centers, which manage subassembly and system assembly operations. At each level, information systems coordinate the manufacturing elements. The CIM system as a whole oversees all the factory's operations, from workstations to cells to centers.

To implement flexible CIM, the factory control systems are supplemented with associated tools and technologies, including simulation models, computer-aided design, computer-aided engineering, group technology, computer-aided process planning, and factory scheduling tools.

Systems management applies information systems at the enterprise level (within or between enterprises) rather than at the level of controlling a specific manufacturing operation (as in CIM). Product data exchange allows business units to exchange computer information generated from their different computer-aided design and computer-aided manufacturing systems. Data-driven management information systems contain the kinds of design, inventory/order, and machine capability information needed to design and manage flexible CIM operations.

Applications Important to the Army

Distributed and forward production facilities consist of manufacturing modules stored together with product subassemblies, raw material, and the electronic knowledge of how to complete the manufacture of finished products to order. The "facilities" are put in place before they are needed. The approach can be applied to simple products, such as clothing, food, and equipment, that can be produced to specific sizes or packaging preference as needed. It is also applicable to larger weapon systems, which can be stored as modules. Upgrades can be made by replacing modules before assembly rather than by retrofitting.

Rapid reaction to operational requirements uses advanced design and manufacturing technology to shorten the cycle from specification to product delivery. For the Army, this application could support specifications sent directly from the field to the manufacturer.

Parts copying uses three-dimensional sensor technology to provide measurements of an existing part, coupled with technologies to etch or sinter raw material to the specified dimensions. Among the potential benefits are storing just one part as a master for copying, making parts without an engineering description, replicating a replacement from pieces of a damaged part, and scaling up or down from one existing size. While this capability is crude and limited at present, it should be applicable to a broad spectrum of parts by 2020.

Issues in Manufacturing Technology

The Advanced Manufacturing Technology Group raised eight issues related to manufacturing for the Army:

- *Sources of Supply.* The current practice of requiring multiple sources of major components or dual-source manufacturing may needlessly increase the cost of defense system acquisition.
- *Material and Component Availability.* U.S. manufacturers are becoming increasingly dependent on foreign sources of basic materials, processes, and components. A number of actions can be taken to ensure that critical materials, skills, and equipment are available if needed by defense forces. Actions may also be needed to limit loss of control over the cost of critical defense materials or systems.
- *Military versus Civilian R&D.* Many of the manufacturing technologies to produce military items will be developed for civilian production first. However, some differences in standards will continue, and some areas of manufacturing will remain unique to the military.
- *Industrial Preparedness.* The ability of the U.S. industrial base to respond to a major mobilization is severely limited. Flexible advanced manufacturing facilities that are capable of rapid conversion from commercial to defense production appear to be the best solution.[6]
- *Capital Investment and Facilities.* Flexible manufacturing systems can help to reduce the investment risk that current procurement practices have placed on industry. Some special contracting arrangements will also be needed. Otherwise, inadequate investment in cost-

---

[6]With respect to conversion from commercial to defense production, international ownership of U.S. manufacturing plants or participation of U.S. facilities in international agreements may complicate the use of these facilities in a wartime emergency. The STAR Committee suggests that long-term agreements be sought to ensure the availability for emergency military use of production facilities in the United States that are foreign owned or controlled.

effective production facilities will only drive up the eventual cost of defense systems.

• *Flexible Production Schedules.* Where the same facilities can serve for both civilian and military production, the Army should consider arrangements for extended delivery schedules to allow military production during off-peak demand for civilian goods.

• *Design for Manufacturability.* Manufacturability should be included as a design evaluation criterion from the outset of requirement formulation. The cost and availability of products will ultimately be driven by the ease and flexibility of production.

• *Environmental and Legal Issues.* Environmental concerns and court decisions about them will continue to affect production facilities of interest to the military. Court decisions like those affecting Department of Energy nuclear material plants will force managers in the government and the private sector to take actions that will affect the cost and availability of defense materials and systems.

ENVIRONMENTAL AND ATMOSPHERIC SCIENCES

TFA Scope

The Technology Group for Environmental and Atmospheric Sciences assessed the following areas of current and projected technology:

• *Terrain-related technologies* include digital topography, terrain imaging sensors, and terrain surface dynamics.

• *Weather-related technologies* include atmospheric sensing, weather modeling and forecasting, modeling of atmospheric transport and diffusion phenomena, and weather modification.

Technology Findings

General Findings

Military operations depend on information about terrain and weather at both large and small scales. As combat operations place increasing emphasis on force mobility and high-technology sensor-dependent systems, the Army will increasingly require a comprehensive information base adequate to:

• characterize operationally significant environmental features (vegetation, soil condition, roads, bodies of water, etc.) of all the land masses of the globe;

- determine how these environmental features are affected by global and local weather patterns; and
- identify variability in "local" weather as a function of locale.

The Army will also need to provide its deployed forces with the capability to sense and interpret the current local conditions of the atmosphere and terrain.

Terrain-Related Technologies

Army units in the field require three categories of information about terrain: topography; environmental features less permanent than topography (roads, amount and type of vegetation, habitation, soil condition); and the capability to anticipate changes in soil condition that may result from weather or enemy action. Topographic data on relatively permanent features can be gathered well in advance of operations and maintained in digital data bases. Advances in digital imagery and in the methods of extracting and storing data from digital images have overcome many of the limitations of two-dimensional maps.

The critical need is for a global, three-dimensional *terrain data base*. Querying and data retrieval must be easy and fast. Yet quick updating of information must also be supported. Techniques are needed to give field commanders dynamic interrogation and viewing of local terrain, plus the ability to generate hard-copy maps for later reference. The enabling technology includes high-capacity opto-electronic storage media, a data base structure for storing three-dimensional data, software and hardware for rapid processing of large data sets, high-speed broadband communications links, multicolor map production from digitized data, microprocessor workstations as the local nodes in this terrain information network, and artificial intelligence to automate reasoning about the interaction of terrain features and other environmental factors, including the weather.[7]

In *terrain sensing* technology, a major breakthrough would be the direct recording of three-dimensional terrain data. An interim evolution is platform-based processing of raw sensor data. Neural network technology can be applied to automated feature extraction. Emerging

---

[7]Near-term efforts in terrain data base development and an advanced concept for a terrain data base system were also considered by the Support Systems Panel. See Volume 3, Part 8.

technology for the identification of features by their wide-spectrum signature (hyperspectral imagery) will also be applicable.

Technology for *real-time terrain analysis* will use computer modeling for which input data from the terrain data base are supplemented with current data from weather and soil sensors. Technological advances will be required in high-resolution terrain sensors, direct observation, and the processing of raw sensor data. Hyperspectral imagery can provide information on subsurface conditions as well as surface characteristics.

Weather-Related Technologies

Weather prediction for operational areas requires atmospheric sensors in the area and the processing capability to synthesize the data into a timely and accurate picture of current conditions, as well as conditions 12 hours to 2 days ahead. The spatial scale for battlefield weather reporting can range from 20 to 200 km.

In the future, *atmospheric sensors* will be flown into forward battlefield areas with UAVs, where they may remain airborne or be dropped to the ground. To provide information on the lowest level of the atmosphere, terrain-following UAVs will be needed. It may also be possible to extract useful weather data from smart weapons. R&D is needed for passive sensing techniques, because the active methods now in use for atmospheric sensing provide targets for the enemy. For remote sensing, satellite LIDAR and radar systems will gather images in wavelengths from the ultraviolet to the microwave region.

With respect to *weather data communications and processing* technology, the multispectral data of the future will require broadband, high-capacity communications links. Data may be relayed to the processing center from local sensors via satellites. To lessen the data transmission load, signal preprocessing in the sensor platform will be applicable. Position location technology will be important in pinpointing the location of sensors transmitting data.

Improvements in civilian-oriented *weather modeling and forecasting* will continue, and the Army will draw upon this technology. In addition, meteorological models are under development for regional use. These have a smaller scale of resolution and rely on sensor data collected on a grid at the same scale. High-speed computers are needed for modeling future conditions from current data and for controlling interpretive displays of both current and predicted conditions. The meteorological community will continue to explore applications of artificial intelligence to weather forecasting. The Army will need to incorporate advances and improve on them for its particular concerns with battlefield scale, effects of combat on local conditions, and weather effects on tactics and equipment performance.

The Army's interest in *atmospheric transport and diffusion modeling* stems from concern with the spread and dilution of CTBW threats, airborne nuclear radiation hazards, pollutants, and battlefield obscurants. Breakthroughs in methods for solving nonlinear stochastic and probability equations for physical, chemical, and meteorological phenomena will allow more realistic modeling of transport and diffusion processes. The projected increase in computing power will also make such modeling more readily available to field commanders.

Even a modest capability to *modify weather* on a local scale, such as lifting fog or initiating precipitation, could have important military consequences. The Technology Group found no particular progress in this area. It concluded that the status of research on weather modification does not merit Army investment at this time, although the Army should continue to monitor this field in case a breakthrough occurs.

4

Advanced Technologies of Importance to the Army

SELECTION OF MOST IMPORTANT TECHNOLOGIES

The Selection Process

The deliberations of the STAR Technology Groups produced a list of more than a hundred technologies with significance for the Army. (Individual technologies at this level of detail are listed in the TFA Scope sections of Chapter 3.) The Science and Technology Subcommittee selected a small number of these as the most likely to produce advances important to ground warfare in the twenty-first century.

The selection was made during a meeting of the Science and Technology Subcommittee in Irvine, California, in April 1990. At a meeting with the Army earlier that year in Washington, D.C., Army representatives had requested a short list of the highest-priority technologies from among those on the full list. The subcommittee group at the Irvine meeting, which included at least one representative from each of the technology groups except Basic Sciences, derived the requested short list by the following process.

The Subcommittee considered the approach of developing specific scenarios for ground warfare. However, this approach seemed too dependent on which scenarios were depicted and even on the details of each scenario. Given the wide range of threat possibilities in the new international environment (see Chapter 1 for details), a more generic picture of future ground warfare was constructed—one that would apply, albeit in a general way, to the gamut of potential threats and conditions under which the Army would face them. This generic picture includes three elements:

- *Attack.* Although the Army has many important roles, an essential mission is to be able to attack and defeat a strong enemy when circumstances require.
- *Defend.* Directly associated with the ability to attack an enemy is the ability to defend itself and U.S. interests (be they civilians, key natural resources, trade routes, the territory of our allies, or the territory of the United States itself) from an enemy's attack. As potential enemies use advanced technologies in their own offensive operations, our defense must stay ahead of them.
- *Gather information.* Information about the enemy's capability, actions, and intentions and about the terrain and weather in which operations will occur has always been crucial to the outcome of ground warfare. The information gathered must be communicated rapidly and often in voluminous detail from the point of initial reception to the nodes in the information network responsible for analysis, integration, decision-making, and action. Finally, those responsible for command and control must be able to understand and act on the information as soon as possible after it has been communicated. Modern warfare, like modern society in general, seems destined to become increasingly dependent on advanced technology for information gathering, processing, and communication. As Chapter 2 pointed out, Operation Desert Storm illustrated just how important winning the information war has become in modern ground warfare.

In considering the relative priority to be assigned to these roles as criteria for the importance of technologies, the Sciences and Technology Subcommittee settled on the following rationale. First, the Army spends most of its time neither attacking nor defending but in a mode of deterrence through readiness to defend and attack. The Army's need for information gathering, communication, and informed decision-making is continuous; it pervades all stages of deterrence, defense, and attack. Second, the traditional and presumably continuing posture of the United States has been to respond to military actions rather than to attack first. Thus, the ability to defend can logically be given second priority, with attack next. Finally, technologies that are key to ensuring that sophisticated technological systems work well and are well integrated with one another should be considered for importance, even if they do not fall into the first rank with respect to information, defense, or attack.

The Subcommittee used these general criteria to evaluate the full list of specific technologies in light of the assessments and forecasts from the technology groups. Each technology group was asked to come up with one, or at most two, candidates, for discussion by the entire subcommittee. From that discussion emerged a list of nine technologies. The following set of nine high-payoff technologies therefore represents a best-judgment selection derived by consensus within a group of about 20 technology experts who had prepared the individual TFAs and had participated in the Subcommittee's deliberations throughout the STAR study.

Selected Technologies

The following nine technologies were selected by the STAR Science and Technology Subcommittee as having the highest priority for the Army:

- technology for multidomain smart sensors;
- terahertz-device electronics;
- secure wideband communications technology;
- battle management software technology;
- solid state lasers and/or coherent diode laser arrays;
- genetically engineered and developed materials and molecules;
- electric-drive technology;
- material formulation techniques for "designer" materials; and
- methods and technology for integrated systems design.

Each of these technologies is described briefly below:

Multidomain smart sensors will be required to locate and target stealthy enemy in camouflage and deception. Passive infrared sensor elements provide information on the angle (direction) of received radiation and the emission intensity. A laser radar sensor element can provide information on reflection intensity, range, range extent, velocity, and angle. Millimeter-wave synthetic aperture radars provide high-resolution doppler images that are responsive to the material properties of targets.

A multidomain sensor incorporating elements such as these can be configured so that the active and passive components share the same optics. This provides pixel-registered images in a multidimensional space, which allows the creation of multidimensional imagery. The richness of the resulting display could give a human observer the capability to detect targets in motion or in concealment under camouflage and trees. Multidomain sensors could also provide high-resolution targeting or act as target designators for remotely launched weapons. Multidimensional smart sensors can be used in counter-stealth systems.

A key area of sensor technology that will contribute to multidomain smart sensors includes *multispectral infrared focal plane arrays* and *uncooled infrared detectors*. Infrared focal plane arrays are the enabling technology for the next generation of night vision equipment. Because they direct more detectors toward the target, focal plane arrays can provide more range and greater sensitivity than previous common-module forward-looking infrared devices. Further development work can improve radiation hardness and spectral bandwidth as well as range and sensitivity. Multispectral (multicolor), wideband focal plane arrays will be needed for robust multimission weapon systems.

Uncooled focal plane arrays do not require cooling with liquid nitrogen, as do current infrared detectors; without the cryogenic cooling, the devices can be lighter and less expensive. For example, this revolutionary technology will allow the Army to expand night vision capability to rifles and weapon sights, passive terminal homing guidance for smart missiles, sights for transport vehicle drivers, and so on.

Improved sensors in smart weapons would reduce ammunition logistics demands, because fewer rounds would be needed to achieve an equivalent effect. Equipment such as a smart helmet for the individual soldier also depends on smart-sensor technology.

The fusion of sensor information by smart processors (derived from model-based or neural network algorithms) could provide the basis for autonomous smart weapons. These may be the best hope for replacing nuclear weapons as the mainstay of defense.

Terahertz electronic devices will be required for increased sensitivity and speed. Electronic devices are the fundamental components of electronic systems such as radar, communications, electronic intercept equipment, and weapon guidance seekers. They are used in front-end receivers and transmitters as preprocessors, as well as in signal processing and automatic target recognition systems.

Today's best electronic devices approach only gigahertz frequencies (a billion cycles per second), but a thousandfold increase in speed to terahertz capability is foreseen. Terahertz electronic devices will be capable of amplifying signals with frequencies as high as a trillion hertz and switching signals at intervals measured in picoseconds (trillionths of a second). These faster devices will make possible much better target identification. Because they are also the building blocks of computers, a great increase in speed from terahertz devices would produce a vast increase in computational power.

Terahertz electronics technology will have the following applications:

- determination of enemy intentions, including likely location of attack,
- surveillance of enemy force movement,
- recognition and identification of enemy forces, and
- guidance of weapons by intelligent seekers.

Secure, wideband communication links are vital for carrying out global army responsibilities. Advanced satellite communications systems will provide sensory access to all parts of the world. However, the complex flow of data from space, air, and ground sensors will require secure high-bandwidth links, even if local preprocessing occurs at the sensor before data are transmitted. Millimeter-wave and optical communication links to satellites, as well as fiber optics networks, offer the greatest potential for secure high-bandwidth transmission for both long distances and local information distribution.

Spread-spectrum electromagnetic links to remotely operated air and ground vehicles will also provide the basis for "telepresence," which enables the intelligence of humans and smart machines to be merged for many applications, including reconnaissance and targeting. The very high bandwidths provided by secure fiber optics systems will permit redundant distribution of sensory and communication information, which is key to robustness in distributed processing.

Battle management software, in the form of a battle control language and associated support, is needed for computer-assisted decision support and battle management. The capacity of computer hardware to process data has increased at a tremendous rate. This capacity is expected to grow by two orders of magnitude every decade. The constraint on fuller use of this capacity is the development of software programs to carry out the types of analysis required for efficient and reliable intelligence extraction, synoptic organization of the intelligence, and interpretation of command decisions into detailed directives to the active elements. For battlefield management, this will continue to be a critical area; it will probably be the pacing factor in implementing an agile-force strategy.

Battle control languages are a layered structure of computer languages. The syntax and semantics of the topmost language duplicate standard military operational and logistical terminology. Statements in this top-level language will look like map graphics, operation orders, or report formats. A series of intermediate languages will provide the ability to modify software at varying levels of abstraction.

Battle control languages will enable Army personnel to move data, extract information, compare courses of action, and even make automated decisions, all without concern for the details of computation. This technology offers capabilities for:

- simulating and evaluating alternative courses of action,
- exercising command and control over the battlefield in near real time with accurate and reliable information, and
- providing an unprecedented degree of realism in training exercises and analytical work.

Among laser technologies of interest to the Army, the two that were judged to have the highest potential payoff were *solid state lasers pumped by diode laser arrays* and *coherent diode laser arrays*. One or both of these technologies could prove valuable for a number of advanced applications.

Solid state lasers based on the rare earth elements and pumped by diode laser arrays are a promising technology for advanced military applications of optics, photonics, and directed energy devices. In contrast to flashlamps, which are the historical method of pumping solid state lasers, diode lasers emit in a narrow spectral band that couples more efficiently into the narrow pump band of the rare earths, delivering the necessary excitation with a much reduced thermal load. This leads to an increase in electrical efficiency of about a factor of 10, with a corresponding reduction in the thermal management needed.

Excessive size and weight for any given performance level have been the major factors inhibiting the use of lasers in military roles; they have been limited to applications requiring only low average power (less than several watts of output power), as in rangefinders and target designators. By relieving the size-performance constraints, diode pumping opens the door to medium- and even high-power applications—up to hundreds of kilowatts. Recent advances in the fabrication technology of diode laser arrays have resulted in cost reductions sufficient to make this approach affordable. Combined with various techniques, both new and old, for wavelength shifting and modulation, the impact of diode pumping on military applications of lasers is likely to be revolutionary. In particular, weapon applications such as antipersonnel weapons, antisensor weapons, and heavy-duty antiaircraft weapons are coming into the realm of practicality. Rangefinders can be expected to expand their performance range to include some search capability and target diagnostics (which may be useful in IFFN). Designators will grow into roles supporting interceptor systems for antisatellite and ballistic missile defense. Diode pumping will also open the door to very compact low-power applications, such as active terminal guidance for projectiles and missiles and target designation for the personal weapon of the individual soldier.

For *coherent diode-laser arrays*, diode laser arrays are coherently coupled to produce the output beam, rather than being used to pump another laser. The Propulsion and Power Technology Group has forecast that power levels of 10 to 1,000 W per modular unit, with energy fluxes as high as 1 kW/cm^2, can be achieved in the next 5 to 10 years. Efficiencies of 50 percent have been forecast. To reach the level of weapon system power (tens to hundreds of kilowatts), this laser technology must be able to extend the mechanisms for phase locking to hundreds or even thousands of diode laser modules into one extended coherent array, while managing the waste heat from high-power operation. Potential applications for coherent diode laser arrays include:

- antisensor weapons to attack enemy surveillance devices and smart weapons;
- antipersonnel weapons;
- infrared illuminators;
- eye-safe and covert rangefinders and other sensors;
- small laser radars (ladars) for special applications, such as motion and vibration sensing;
- sensors for battlefield IFFN; and
- line-of-sight wideband communications.

Genetically engineered and developed materials and molecules. Within 15 to 30 years, biosensors derived from the human immune system will provide early warning of chemical, toxin, and biological warfare (CTBW) agents. Soldiers will be immunologically enhanced for global protection from naturally occurring endemic infectious disease organisms, which will probably remain the largest casualty producer in future combat situations. Expert medical diagnostic systems in palm-top computers will allow nonspecialist personnel to make rapid diagnoses. Using very rapid recombinant DNA technology, disease organisms can then be isolated and specific vaccines produced within days.

Biotechnology will be able to produce both natural and artificial materials—such as composites and customized polymers—with specified physical, chemical, and electrical properties. Advances in this area will depend on the simultaneous development of computer-aided biomolecular design and low-temperature manufacturing. These are some of the potential implications of materials produced with biotechnology:

- increasing the number of effective personnel in battle by speeding the return to duty of injured soldiers;
- providing greater troop mobility by defeating CTBW barriers;
- using unmanned sensors (UAVs or UGVs) carrying CTBW biosensor systems to detect the presence of CTBW threats before troops move into an area; and
- providing biologically derived aerosols and other "soft kill" weapons to defeat enemy vehicles by causing engine malfunction.

Electric-drive technology can increase battlefield mobility and effectiveness of ground vehicles. Integrated propulsion systems that combine electric drives with advanced diesel or gas turbine engines for primary power offer major gains in total battle zone effectiveness and mobility. Combining an advanced engine with an advanced electric drive that distributes power flexibility to each wheel or track will significantly improve power density and weight distribution while decreasing signatures and fuel consumption. Power plant options for these integrated systems include ultra-high-temperature quasi-stoichiometric gas turbines with high-pressure ratios and nonrecuperative simple cycles.

The basic modules of an integrated electric-drive system could be used interchangeably among different vehicle types or in other battle zone systems of a highly "electrified" Army of the future. Power distribution systems for mobile platforms based on electrical power would complement the electric-powered weapons in the battlefield of the future. The benefits of this technology include (1) greater vehicle design flexibility; (2) greater power density and vehicle mobility; (3) reduced fuel consumption, which will reduce the logistics burden; and (4) integration of electric-powered weapons with the vehicle propulsion system.

Advanced material formulation techniques and advanced materials will provide specific properties to satisfy the performance requirements of future systems. Conventional primary materials such as steel, aluminum, or titanium cannot provide the combination of properties required for many advanced Army requirements. For example, the success of advanced rail gun concepts may well hinge on developing hybrid or multifunctional materials that simultaneously provide extremely high values for electrical conductivity, wear resistance, and stiffness. Only tailored macrocomposites (combining advanced metallic, polymeric, and ceramic materials) are likely to provide the requisite combination of structural, optical, and electronic properties.

Integrated system design technologies can dramatically lower system costs, shorten development cycles, and enhance system effectiveness, reliability, and flexibility. Concurrent development methods, coupled with computer-based design environments and applications of artificial intelligence, provide a powerful new tool for developing optimal designs rapidly. Set-based inference systems allow the development of low-cost systems that will accommodate a wide range of manufacturing and environmental variations.

High-level representation languages, with associated compilers, already speed very-large-scale electronic circuit integration (VLSI) and software design. They are beginning to be applied to the design of complex mechanical systems as well. Simulation and fast prototyping techniques allow quick and early checks on design feasibility. Newly evolving methods for managing the design process can replace rigid and arbitrary specifications with the simultaneous and systematic exploration of alternatives and trade-offs among doctrine, product, and manufacturing process. Working synergistically over the next 30 years, these technologies are expected to lower costs while enabling swifter, more reliable, and more flexible development.

GENERAL CONCLUSIONS FROM
THE SELECTION PROCESS

As the Science and Technology Subcommittee deliberated on its selection of a short list of high-payoff technologies, its members were led to several general conclusions that the Subcommittee thought will have as much significance to the Army as any list of a few particular technologies. The STAR Committee believes that these conclusions, drawn by a representative body of the STAR study's technology experts, are important enough to the major themes of this main report that they bear repeating. Parenthetical comments have been added to indicate where the theme of each conclusion is further elaborated.

• The foreseeable evolution of technology will profoundly affect warfare. (Chapter 6 elaborates some of the prospects for long-term implications of technological changes on force structure and strategy.)

• There are so many technologies with important military consequences that a primary problem for Army technology management will be to select focal interests and implement them effectively. (Chapter 5 addresses the need for a technology implementation strategy with a defined set of focal interests.)

• Many important technological advances are occurring outside the United States. The Army will need to consider how to make effective use of technology from worldwide sources.

• Advanced military technologies will be widely available throughout the world. The Army will need the means to deal with well-equipped and sophisticated enemies, even in smaller conflicts. (The section of Chapter 6 on near-term force structure implications addresses this issue.)

ENABLING TECHNOLOGIES FOR NOTIONAL SYSTEMS

Although the selection of high-payoff technologies addresses the Army's request for a short list of the highest priorities, there is much significant technology that any such list must omit. Figure 4-1 summarizes the relevance of the broader range of advanced technologies discussed in Chapter 3 to the notional systems envisioned by the eight STAR systems panels.

The columns of Figure 4-1 were selected from among the key notional systems described briefly in Chapter 2 and more fully in the systems panels' reports. The rows of this matrix are classified under headings corresponding to the eight technology group reports. The individual rows under a heading correspond to subheadings used in Chapter 3, which also reflect the topical structure of the technology groups' reports.

The symbols entered in this matrix indicate the STAR Committee's assessment of the degree of relevance the technology (or area) has to a notional system (or class of functionally related systems). This assessment is based primarily on the findings of the systems panels and technology groups. Figure 4-1 is not intended as a definitive statement of the importance of specific technologies to different systems. Its primary purpose is to serve as a reader's guide to the more complete presentations found in the systems panels' reports and the Technology Forecast Assessments.

COMPARISON WITH OTHER TECHNOLOGY LISTS

Appendix A compares the STAR lists of high-payoff technologies and high-payoff notional systems with three other lists: (1) the Army Technology Base Master Plan, (2) the Defense Critical Technologies List prepared by the Department of Defense, and (3) the list presented in the *Report of the National Critical Technologies Panel.*

● Advances Required
⊗ Advances Important
○ Advances Relevant

BATTLEFIELD FUNCTIONS

TECHNOLOGIES / SYSTEMS CONCEPTS	Information War						Integrated Soldier Support					
	Electro-Optical Sensors	UAV-Borne Sensor Systems	UGVs for C3I/RISTA	C3I/RISTA Network	Electronic Sys. Architect.	Space-Based Systems	Combat Systems	Support Systems	Disease Prevent. & Treat.	Battlefield Preventive Med.	Battlefield Injury Treatment	Robotic Helper Systems
Comp. Sci., Art. Intell., Robotics												
Integrated System Develop.		○	●	●	●	●	○					○
Knowledge Rep. & Languages		○	○	●	⊗	⊗		●				⊗
Network Management	○	●	⊗	●	●	●		○				
Distributed Processing	○			●	●	●		⊗				
Human-Machine Interfaces		⊗	⊗	●	●	○	⊗	●				⊗
Battlefield Robotics		●	●				○	○				●
Technologies to Monitor	○	⊗	⊗	⊗	⊗	○		⊗	●		⊗	⊗
Electronics and Sensors												
Electronic Devices	○	⊗	●	⊗	○	⊗	○	⊗				○
Data Processors	○	●	●	●		●						○
Communication Systems		●	⊗	⊗	○	●		⊗				
Sensor Systems	●	●	●	●		⊗	○					○
Optics, Photonics, Dir. Energy												
Optical Sensor & Display Tech.	●	⊗	●	○			⊗	○				
Photonics & EO Technology	●	⊗	⊗	⊗		⊗		○				
Directed Energy Devices							⊗					
Biotechnology & Biochemistry												
Gene Technologies							●	●	●	●		
Biomolecular Engineering				○			●	●	●	●		
Bioproduction Technologies							⊗					
Targeted Delivery Systems	○						●	●				
Biocoupling							⊗					
Bionics	○			○			○				●	
Advanced Materials												
Resin Matrix composites			●			○					●	
Ceramics			○									
Metals												
Energetic Materials												

TECHNOLOGIES / SYSTEMS CONCEPTS	Combat Power & Mobility										A& MD	Combat Service Support								
	Vehicle Drive Systems	Road Building & Bridging	Helicopter & Transport UAVs	Advanced Armored Vehicle	Anti-Armor	Brilliant Munitions	Lt. Wt. Indirect Fire	Directed Energy Weapons	Mine & Counter-Mine Opns.	Airborne Ground Attack	Air and Missile Defense	Health & Medical	Mapping Systems	Shelter	Ammunitions Support	Fuel	Maintenance and Repair	Support C3	Simulation Systems	Training & Personnel
Comp. Sci., Art. Intell., Robotics																				
Integrated System Develop.	●		○			⊗	●	⊗		⊗	●								●	○
Knowledge Rep. & Languages											●									⊗
Network Management										⊗	●						⊗	⊗	●	⊗
Distributed Processing						○					●		○				⊗	⊗	●	⊗
Human-Machine Interfaces			○				○			○	⊗	○	○				○	○	●	●
Battlefield Robotics		⊗	⊗			⊗			●	⊗	●						⊗			○
Technologies to Monitor	○					●	⊗	○	○	○	●	⊗	○				⊗	●	●	○
Electronics and Sensors																				
Electronic Devices	○					●	○	⊗		⊗	●	●								
Data Processors						●	○	⊗		⊗	●	●					○			⊗
Communication Systems						●	○			●	●	●					○			
Sensor Systems						●		⊗	⊗	●	●	●	○							
Optics, Photonics, Dir. Energy																				
Optical Sensor & Display Tech.						●		○		●	⊗		○				○			
Photonics & EO Technology						●		⊗		○	⊗									
Directed Energy Devices					○	⊗		●	○	○	⊗									
Biotechnology & Biochemistry																				
Gene Technologies									○			⊗								
Biomolecular Engineering												⊗								
Bioproduction Technologies												○								
Targeted Delivery Systems									○			⊗								
Biocoupling												○								
Bionics																				
Advanced Materials																				
Resin Matrix composites	●		○	●	○					⊗				⊗			⊗			
Ceramics	●			⊗	○					○				⊗			⊗			
Metals		○	⊗	⊗	●					⊗				⊗			⊗			
Energetic Materials				○	●				⊗	⊗	○				⊗					

BATTLEFIELD FUNCTIONS

Information War · Integrated Soldier Support · Combat Power & Mobility · A& MD · Combat Service Support

SYSTEMS CONCEPTS

Electro-Optical Sensors · UAV-Borne Sensor Systems · UGVs for C3I/RISTA · C3I/RISTA Network · Electronic Sys. Architect. · Space-Based Systems · Combat Systems · Support Systems · Disease Prevent. & Treat. · Battlefield Preventive Med. · Robotic Helper Systems · Vehicle Drive Systems · Road Building & Bridging · Helicopter & Transport UAVs · Advanced Armored Vehicle · Anti-Armor · Brilliant Munitions · Lt. Wt. Indirect Fire · Directed Energy Weapons · Mine & Counter-Mine Ops. · Airborne Ground Attack · Air and Missile Defense · Health & Medical · Mapping Systems · Shelter · Ammunitions Support · Fuel · Maintenance and Repair · Support C3 · Simulation Systems · Training & Personnel

TECHNOLOGIES

Propulsion and Power
- High-Energy Lasers
- RF, Microwave, & MM Wave
- Missile Propulsion
- Air Vehicle Propulsion
- Surface Mobility Propulsion
- Gun/Tube Projectile Propulsion
- Battle Zone Electric Power

Manufacturing Technologies
- Designer Parts
- Distrib. & Forward Production
- Rapid Response to Field Reqs.
- Parts Copying

Environ. & Atmos. Science
- Terrain-Related Technologies
- Weather Modeling & Forecasting
- Weather Modification

Long-Term Forecast Trends
- Information Explosion
- Computer Simulation/Visualization
- Control of Nanoscale Processes
- Chemical Synthesis by Design
- Complex Systems Design
- Materials Design by Computation
- Hybrid Materials
- Advanced Manufact. & Process.
- Biomolecular Structure & Function
- Biological Information Processing
- Environmental Protection

Legend:
- ● Advances Required
- ⊗ Advances Important
- ○ Advances Relevant

FIGURE 4-1 Relevance of STAR technologies to representative systems concepts.

5

Technology
Management Strategy

INTRODUCTION

This chapter addresses the second point in the STAR request. It suggests a technology management strategy to realize the potential offered by the technology applications discussed in the previous three chapters. In preparing this strategy, the STAR Committee drew extensively on the report by the STAR Technology Management and Development Planning (TMDP) Subcommittee; the complete report is available as a separate volume. Some of the reports of the systems panels and technology groups also contain discussion and recommendations on issues of technology management. The discussion here draws selectively on these reports, which the STAR Committee interpreted within its own perspective. Interested readers are urged to consult the separate reports for additional detail and, in some instances, different emphases and opinions.

The STAR Committee's suggestions to the Army on technology management are organized here under five headings:

- a general *implementation strategy* to move new technology into Army systems;
- a set of recommended *focal values* for technology implementation, which apply across the major combat and support functions and can serve as key elements of a strategic focus on advanced technology applications;
- technology management recommendations specific to each of the *high-impact functions* discussed in Chapter 2;

- recommendations for enhancing the Army's *in-house R&D infrastructure*; and
- an evaluation of the current Army *requirements process* as it affects technology management, with recommendations on how to improve it.

IMPLEMENTATION STRATEGY

Figure 5-1 illustrates how an *implementation policy* can be used to achieve a *strategic focus* for technology management. This section begins with a general implementation policy and a general statement of the basis for focal interests. The remainder of the section recommends specific actions to carry out the implementation policy. These implementation actions are summarized in the list on the left side of Figure 5-1.

The next major section, Focal Values, recommends seven aspects of systems that should be emphasized as part of the Army's strategic focus on technology management. The third section, High-Impact Functions, applies the implementation strategy and the focal values to technology management issues within each of the principal combat and support functions discussed in Chapter 2. The focal values and major topics discussed in the third section are listed on the right side of Figure 5-1.

General Statements

Implementation Policy. The STAR Committee recommends that the Army direct most of its available resources toward those technologies and applications that are not receiving sufficient private sector investment to meet anticipated Army interest. The high-payoff technologies identified in Chapter 4 and the high-payoff systems listed in Chapter 2 represent the Committee's nominations for initial lists of such technologies and systems. Furthermore, the Army should, wherever possible, increase its reliance on the private sector for technological progress and products.

After the Army has considered the implementation actions and focal interests recommended below, it will probably revise or delete some of them, while adding others of its own choosing. Although the STAR Committee believes each of its recommendations to be well justified, more important than any specific recommendation is a focused implementation strategy. That strategy must have a clear set of focal interests, and these must be implemented through actions that can be communicated throughout the organization.

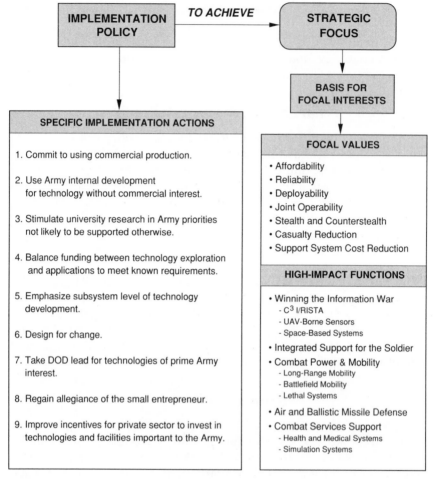

FIGURE 5-1   A focused implementation strategy for technology management.

Basis for Focal Interests. The Army should focus its technology development toward explicit Army system interests as a means of exploiting advanced technology more fully and of transferring new technology more rapidly to the field. These Army focal interests should fit within the larger defense policy architecture of the Office of the Secretary of Defense (OSD). They should be explicitly supportable by reference to that architecture.

Some of these focal interests will apply across many systems, like the seven focal values recommended in the next section. Other focal interests will be specific to a systems concept, such as those recommended in the third section of this chapter. The Army's implementation of technology will progress more rapidly and cost-effectively if those responsible for it can envision an important application requiring that technology. In Army-sponsored research and exploratory development, whether performed at universities or in Army laboratories, a reasonable balance must be struck between unrestricted freedom to explore and the discipline imposed by specific applications.

To have its acquisition funding requirements understood and accepted at higher levels, the Army must articulate its focal interests by reference to the larger priorities set forth by OSD. Two key issues must be addressed with more consistency and depth than has sometimes occurred in the past:

• Is the Army in fact working on technologies and systems that fit into the wider architecture of OSD priorities?
• Where the Army is working within that architecture, are the arguments for support being couched in the context of OSD priorities or are they tied solely to Army-specific interests?

Implementation Actions

Nine of the recommendations from the TMDP Subcommittee are presented here as key elements of an implementation strategy for technology management. The TMDP Subcommittee report amplifies the summary account of these implementation elements given here.

• Commit to using commercial technologies, products, and production capabilities wherever they can be adapted to meet Army needs. The STAR Committee concurs with the argument made by the Defense Science Board and endorsed by the TMDP Subcommittee that the issue here is not essentially one of cost. Rather, DOD will have neither the resources nor the production volume to support on its own an expensive, dynamic infrastructure for advanced technology manufacturing. A commitment to use commercial products lets the Army benefit from market-driven and market-financed technological developments. It also provides timely surge capacity from existing private sector manufacturing facilities; surge capacity will almost certainly be required in future contingency operations, as it was in Desert Storm.
• Focus the Army's internal technology R&D in areas where strong private sector interest is not anticipated. A special section of this chapter (see below) deals specifically with Army R&D infrastructure;

there is additional valuable comment in the TMDP Subcommittee report. These STAR discussions are directed not at *what* the Army should be researching and developing in-house but at *how* internal R&D can be managed to achieve long-term results in whatever technological or systems areas are selected.

• Stimulate university research in technologies important to the Army that are not likely to receive adequate support either from the private sector or through other grant mechanisms.

• Balance technology funding between the exploration of new concepts made possible by scientific advances and the specific technological applications needed for Army systems. In striking this balance, the Army should consider whether commercially developed technologies and products that meet military needs will be available. To encourage leap-frog technological solutions, innovative research and design must be encouraged. Still, a clear focus on the functional characteristics of a system can help guide this innovation. It is also important to foster an environment of intellectual freedom and challenge in the Army's laboratories and research centers, even while R&D managers are held accountable for productivity.

Overall, the STAR Committee concludes that a strategic focus on Army interests ought to be maintained. This conclusion is embodied in the general statement on strategic focus.

• To modernize the current inventory, the Army should pay more attention to the subsystem level. Technology development, operational demonstration, and production proofing can be applied to subsystems; these can be upgraded if the larger system or platform has been designed for change. The TMDP Subcommittee has presented cogent arguments for combining an increased focus on subsystems with new approaches to platform development. This six-point plan could significantly shorten the platform development cycle and reduce costs. The same points were made in the 1984 Defense Science Board Summer Study on upgrading current equipment.

• Design systems to accommodate change during the design life of a system. As a consequence of both budget constraints and increasing technical complexity, major platforms are likely to have longer *generational* cycles (as opposed to the development cycle discussed above). To maintain a technological advantage, the Army must be able to modernize without waiting for the next generation of an entire system. However, successful retrofitting of an integrated system requires that it be designed initially for modular replacement of components and subsystems as they are upgraded.

• Seek to become the DOD lead agent for technologies of prime interest to the Army; consider taking on roles in other DOD programs as a means of ensuring DOD activity in areas of broadly useful technology. Technologies for theater-level command and control and for training individuals or units are good examples.

• Revise Army procedures and practices to provide incentives for entrepreneurial small businesses to contract with the Army. The TMDP Subcommittee has graphically portrayed the alienation of this innovative segment of the private sector from the Army market. To recover the situation, changes are needed in progress payments, cost of competitive procurements, intellectual property rights, and other areas.

• Improve incentives for the private sector to invest in DOD-unique technologies, applications, and specialized facilities. Profit policies and amortization requirements are examples of areas where the Army could press for changes in legislation or DOD directives.

FOCAL VALUES

As noted in the introduction to Chapter 2, the STAR Committee found a number of key values that were cited repeatedly by the STAR systems panels and technology groups as potential benefits of many emerging technologies or systems concepts. These same characteristics were selected by senior military leaders or advisers as important to the success of the Army in its anticipated future roles. The STAR Committee has selected the seven most pervasive and potentially beneficial of these attributes to recommend as focal values for the Army's technology implementation strategy (Figure 5-2). How technology can contribute to each of these values will, of course, differ from system to system.

Affordability

The STAR Committee believes that the traditional Army Technology Base Program puts far too little emphasis on the application of technology expressly to achieve affordability. Individual short-term product acquisition programs cannot be expected to invest heavily in technology-based affordability initiatives if, because of development cycle timing, the savings attained cannot accrue benefits for that program. The Army should allocate a significant portion of its research, exploratory development, and advanced development resources[1] for

---

[1]These three areas correspond to budget lines 6.1, 6.2, and 6.3a, respectively.

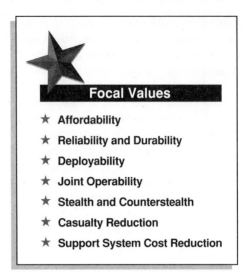

FIGURE 5-2 These seven technology management values apply across Army systems and technology development efforts in all mission areas.

a sustained program focus on cost reduction opportunities from advanced technologies. Furthermore, the Army should pursue, through both its own laboratory system and the private sector, a broad investigation of technology options that show potential for major cost reductions.

Several Army laboratories are already leading DOD in this important area of a technology focus on affordability. This effort should be substantially expanded, in light of the budget restrictions anticipated during the next decades.

The Army should consider ways to stimulate industry investment in flexible manufacturing systems. These systems have promise as a means of economical production even at low and fluctuating rates, which is likely to be the pattern of much future defense manufacturing demand. Similar manufacturing technology will be needed in the Army's own facilities, most importantly its depots.

Reliability

More reliable systems can achieve improved performance and reduce costs at the same time (Figure 5-3). In the field, reliability becomes essential to maintaining and exercising technological advantage. While these benefits of more reliable systems are undisputed, the means of improving reliability are not always obvious. The STAR Committee sees a number of technology developments, driven by competition in private sector markets for more reliable products and services, that should be tapped for exploitation by the Army. Two such areas are low-failure electronic and electromechanical systems and improved software producibility.

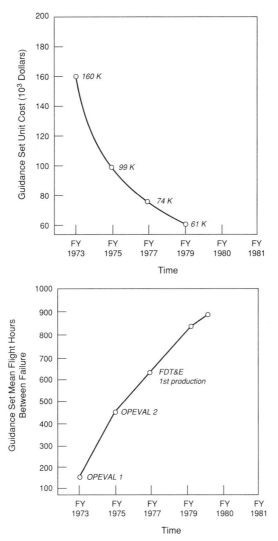

FIGURE 5-3 Experience with the Navy's Sparrow missile shows that reliability can be increased over a system's lifecycle, even while unit costs are decreased. (SOURCE: China Lake Naval Air Station.)

Low-Failure Electronic and Electromechanical Systems

Production management with the goal of improving product reliability is becoming increasingly important in highly competitive commercial markets, including consumer electronics (televisions, video cassette recorders, and personal computers), automobiles, and other areas where the consumer buys high-tech goods. The technology to improve product reliability includes manufacturing methods, failure-mode sensing techniques, and new materials that wear gradually with observable symptoms ("graceful" failure) rather than failing all at once (catastrophic failure). In electronics, new semiconductor materials, new device formation techniques, and electronic design automation are all contributing to improved reliability. For applications ranging from micro-devices to large-scale mechanical systems—such as engines and bearings—and structural components, advanced materials can now be designed, produced, and fabricated to address sources of failure inherent in older materials or production methods.

One way for the Army to exploit this "reliability revolution" in the marketplace is to rely more on proven commercial products and components. In other instances the techniques, materials, or methods can be adapted for use in Army-unique products.

Improved Software Producibility

Software engineers have begun to codify software development principles and practices that are essential when computer programs contain millions of lines of code. They already use practices such as structured programming, rapid prototyping, and software reuse, with the aim of helping software performance keep up with advances in computer hardware performance. However, the large software systems made possible by future hardware will require software engineering techniques far more advanced than the current state of the art. In particular, the verification and validation of mammoth software systems present a daunting challenge, for which an integrated software technology program is necessary.

To participate in this evolving field, the Army should institute a programwide focus on software producibility aimed at tracking progress in software development methodologies and applying them to Army systems. The STAR working groups found that considerable research on software tools and environments is being sponsored by DARPA, the Office of Naval Research, and the National Science Foundation. In addition, the DOD Software Engineering Institute and most universities have active software research programs. So do many private companies, including consortia formed by hardware manufacturers. There is an extraordinary level of activity in the private sector, both domestic and foreign, which can the Army can use if it organizes to do so.

Deployability

In a future where the Army must be prepared to respond quickly to sudden contingencies anywhere in the world with forces based primarily in the continental United States, deployability takes on new meanings and new urgency. It introduces new constraints on the formulation of requirements and the design of systems; one can no longer divorce the military effectiveness of a system from the time and manner in which it can be transported halfway around the globe. In addition to long-distance transport into theater, mobility within theater also must be considered. The existence of a modern infrastructure of highways, railroads, and airfields cannot be assumed. In addition to the transport and battle zone mobility of combat troops and their systems, deployability also applies to the logistical support of forces once they are in the field.

Fortunately, the technologies investigated by the STAR panels offer myriad possibilities for enhancing deployability in each of these aspects. More compact systems can be developed, using lighter materials. Equal or greater lethal power can be "packaged" in smaller, lighter systems. New materials, custom designed for specific applications, offer more strength and toughness in lighter, thinner formulations. Smart munitions in small quantities can achieve the same destructive effect as tons of conventionally delivered explosives. The challenge will be to ferret out the best options for a given requirement and test them adequately to make the right choice, within the limited resources available.

Joint Operability

Although multiservice combat operations have been the norm in the past, in future contingencies the complexities of such operations will increase. At a systems level, future Army weapon and information systems will need to work together with the analogous systems of the other services. During Desert Storm the services managed coordinated operation of systems that had been designed for use by each service operating independently. However, to achieve systems that will later work well together, decisions on a multiservice architecture and standardized key components will have to be made by each service early in the development cycle of its systems.

Army management of the technology programs must maintain a focus on this joint operability environment. In particular, the Army will need to cooperate closely with the other services on programs for the interoperable systems with high payoff, such as air and ballistic missile defense or C³I/RISTA (command, control, communication, and intelligence or reconnaissance, intelligence, surveillance, and target acquisition) systems. Specific examples of technology in these areas that will require close cooperation are jamming and IFFN (identification of friend, foe, or neutral). Special consideration must be given to the selection of time and frequency domains, power levels, and interfaces for both voice and data interchange.

Stealth and Counterstealth Capability

In addition to applying low-observable (LO) technologies to its weapon systems and its airborne and ground vehicles, the Army should apply stealth technologies to its logistical support facilities and equipment. The application of stealth technologies to these facilities and equipment can make them much more difficult for the enemy to locate, target, and attack. In the contingency environment of the future, these additional means of protecting logistics bases could become critically important to mission success.

The STAR working groups identified a variety of promising technologies for advanced materials and for the reduction of electronic, optical, and sound signatures. The STAR Committee suggests a managed focus on systems applications to integrate and direct these still-emerging technologies.

The Army is not yet totally prepared to defend itself against stealth weapons. As these technologies become more widely available, countermeasures to them will gain in importance. Particularly in rapid-deployment contingency operations, fixed concentrations of forces and logistic lodgements are likely to be far more vulnerable to the level of stealth technology available to an adversary than they have been in recent operations. At present, U.S. forces enjoy an overwhelming advantage in air superiority and the use of stand-off ground weapons. This important technological advantage could be eroded if an adversary can threaten U.S. forces, even if to only a limited degree.

The detection of unfriendly stealthy systems by means of advanced radar and other sensor technologies appears plausible. The STAR Committee recommends that the Army take the lead in an extensive multiservice program for an integrated approach to the detection of stealthy attack.

Casualty Reduction

The focal value of casualty reduction is intended to apply to a range of overlapping concerns. Foremost among these is the prevention of injury or death to U.S. or friendly forces in combat. But casualty reduction also encompasses prevention or amelioration of (1) lost troop strength from non-life-threatening diseases, noncombat injuries, or inadequate protection in harsh environments; (2) disability, pain, and psychological effects from serious injuries, whether or not combat-related, and from life-threatening diseases; and (3) deaths or injuries sustained by noncombatants caught in the battle zone.

Battlefield medicine and the other health and medical systems concepts described in Chapter 2 are obviously key means to casualty reduction in all these senses. Armor for manned vehicles, ballistic protection for the individual soldier (helmet and special clothing), and unambiguous IFFN also reflect this pervasive value. Less direct, but important nonetheless, are the casualty reduction consequences of training technologies, substitution of unmanned systems for manned systems in hazardous roles, and improvement of soldiers' shelter and rations.

Support Systems Cost Reduction

In close alliance with the values of affordability and reliability, the Army should institute an across-the-board technology management goal to reduce support system costs. Reduction in the cost of both the systems themselves and the support manpower they require is essential to the leaner Army of the future. Among the technologies assessed by the STAR working groups that can aid in this cause are low-failure electronics and improvements in the expected lifetime of mechanical systems. Mechanical hardware designed for minimum maintenance can greatly reduce support systems infrastructure and cost. Also, modern methods of reduced cost maintenance are applying technological advances in materials, software engineering, simulation, testing, and manufacturing.

Within the private sector, major programs to put these technologies into practice are under way; the Army can profitably follow the private sector's lead. The Army can also learn from the private sector's successes in technology-intensive systems for large-scale inventory control and distribution.

With respect to reducing maintenance costs, a goal of "mean time between removals" of components or subsystems must be considered along with the goal of increasing the "mean time between failures." Complex systems, such as integrated electronics suites, tend to have a high incidence of false removals; the time and stockage required for total removals, not just failures, drives maintenance costs. For this reason, reducing maintenance costs must also focus on fault detection and isolation techniques, such as embedded sensors in "smart materials" or embedded diagnostics in software systems. Improved fault detection and isolation for highly integrated systems can substantially reduce the maintenance burden of fault diagnosis; it should be a key design requirement of such systems.

FOCAL INTERESTS WITHIN
THE HIGH-IMPACT FUNCTIONS

The preceding sections of this chapter recommended an implementation strategy for technology management and an initial set of pervasive values on which that strategy could focus. This section applies the strategy and focal values to advanced systems concepts for each of the high-impact functions discussed in Chapter 2.

The STAR Committee recommends that the Army formally designate a group of high-impact functional areas into which most of its technology exploitation effort should be channeled. An illustration is the recent Soldier-as-a-System program. Although this report and the Army Science Board have made recommendations to improve this program, it is nonetheless a good example of a technology focus on an important Army functional area.

Winning the Information War

C³I/RISTA

The emerging technologies assessed by the STAR working groups will make possible future Army C³I/RISTA systems of vastly improved performance. But to achieve this critical capability, the Army must use management methods that accelerate and concentrate the application of the enabling technologies to the uniquely military requirements of C³I/RISTA.

Critical C³I/RISTA applications with uniquely military requirements include:

- IFFN;
- advanced multisensor integration;

- situation assessment and alternative approach evaluation through advanced hardware and software computational techniques;
- automated remote sensing, both ground based and airborne; and
- target acquisition in extremely difficult environments.

The Army should also exploit relevant developments in the private sector. For example, the private sector is rapidly moving to an environment in which multimedia communications will be widely networked through new and affordable telecommunications techniques. These commercial information handling systems of the near future will be able to assemble, sort, fuse, and disseminate immense quantities of diverse information in clear and flexible ways. The Army should prepare now to make the fullest and most timely use of these communications developments for next-generation C^3I/RISTA.

The private sector will soon be able to explore "what if" scenarios in real time, with sufficient complexity and realism to train, and eventually to aid, managers at all levels of operational decision-making. In the longer term, emerging technology should produce expert systems capable of evaluating trend data in real time, while the consequences of decisions are still nascent and timely changes can still alter operational outcomes for the better. The potential in these decision-support technologies for command-and-control applications is easily imagined. But realization of that potential will require both receptivity to commercial technology from the private sector and its successful adaptation to the military world of C^3I/RISTA.

UAV-Borne Sensor Systems

The achievements of UAVs have thus far fallen short of what was predicted by previous panels of the Army Science Board and Defense Science Board, and UAVs have never received full acceptance by the Army. The STAR Committee believes, nonetheless, that major causes of this failure of UAVs or RPVs (remotely piloted vehicles) have been fragmented development programs and a lack of clear management focus on the capabilities to be implemented.

The Committee still believes that UAVs have remarkable potential for a spectrum of Army C^3I/RISTA applications. At one extreme, small, special-purpose machines could support dismounted soldiers. At the other, large, multipurpose systems, like the high-altitude, long-endurance (HALE) system considered by the Airborne Systems Panel, could provide information critical to corps and division commanders. Thus, the Committee urges the Army, as matter of high priority, to organize the diverse interests that would benefit from this technology into a constituency for an integrated technology development program in UAV applications.

Space-Based Systems

Space-based systems, which include reconnaissance, early warning, and communications satellites, will become increasingly essential elements of C^3I/RISTA systems. During the Desert Shield and Desert Storm operations, for example, the U.S. Army Central Command apparently[2] used satellites for communications, location and maneuver, terrain mapping, environmental assessments and prediction, ballistic missile early warning, and battle damage assessment. To realize more of these benefits in support of its mission, the Army must make a commitment to long-term dependence on, and support for, space-based systems.

A key issue, which was debated but not fully resolved within the STAR Committee, concerns the extent to which the Army should rely on national systems and systems of the other services as opposed to investing in Army-dedicated space-based assets. There was general agreement that the overall Army strategy should be to use national assets or assets developed by other services and agencies when there is opportunity to do so without risk to Army needs. In periods of crisis and warfighting, however, space communications and RISTA assets that were previously designated for Army use have been diverted to other tasks by national command levels higher than the Army. This pattern, which occurred again in Desert Storm, is the rationale for a limited system of Army-dedicated satellites, in particular for functions that require Army control of the uplink. For downlink-only applications, such as location and maneuver, terrain mapping, or weather information, reliance on "someone else's" satellites poses no problem.

The argument against Army investment in developing and maintaining its own assets was presented not merely on the basis of cost (which would be significant) but, more importantly, in terms of keeping the Army's efforts focused on its areas of competency. Other services (notably, the Air Force and the Navy) and agencies (NASA), as well as the private sector (e.g., telecommunications companies),

---

[2]The STAR Committee was not briefed by the Army on these operations, but available information indicates that space-based systems were used for the functions listed here.

have special competency in space-based systems and are committed to the pursuit of technological advances. It would therefore be wiser, from this position, to make use of competencies that others possess rather than allocating scarce resources to an undertaking likely to result in second-rate capabilities.

A final resolution of these different views can emerge over time; in any case, the matter is not something the STAR Committee can or should decide. The Committee recommends that, at the least, the Army dedicate the resources of personnel and technical capabilities needed to become an active and vocal participant with the other services and elements of DOD in planning and operating future space-based systems. From within this framework of active participation and improved understanding of the options, the Army will be in a better position to determine how far it can rely on someone else's assets and how best to exploit the technological capabilities of other players to fulfill Army requirements.

Integrated Support for the Soldier

Army policy is already placing increased emphasis on the soldier, and the trends in technology support this emphasis. Viewing the individual soldier from the perspective of systems analysis, as recommended in Chapter 2, requires integration of many hard science specialties, such as those for hardware and software, as well as competent use of advances in understanding human performance. The STAR Committee recommends that the Army's current Soldier-as-a-System program become the starting point of a much broader initiative for integrated development of technologies to support the soldier in many roles, not just the dismounted foot soldier. The STAR Committee endorses the recommendation of the Special Technologies and Systems Panel for establishment of a Soldier Systems Research, Development, and Engineering Center. This center would conduct programs needed to implement key emerging technologies for integrated support of the soldier. It would also maintain a technology watch for innovative ideas and applications.

Particularly in the hardware aspects of soldier-oriented technology, a systems approach to the core capability is required. This approach should include new combat protection and capabilities for detection, sensing, communications, and offensive operations. Developments in these areas should be phased into a modular architecture as they become available. The architecture should allow for a range of options, with selection among them tailored to the particular assignment of the soldier.

With respect to mobility, the dismounted soldier should not be constrained by the load that can be carried when on foot or when parachuting onto the battlefield. The STAR Committee leans toward the robot "mastiff/mule" concepts rather than an exoskeleton approach. But whatever systems concept is decided on, some form of electromechanical assistance is needed. Furthermore, it must be consistent with the environment of the modern battlefield. For example, it cannot have acoustic, heat, or other signatures that drastically increase the risk of detection and targeting by the enemy.

Parachute systems and practices are another area of soldier mobility that requires a systems approach to load management and technology. The Special Technologies and Systems Panel reviewed data from Operation Just Cause in which jump injuries were a major cause of casualties requiring evacuation out of theater (20 percent by one report). Among the contributing factors cited were jump loads that were too heavy to be properly lowered (released) given the low altitude and uncertain terrain of the operation. Similar conditions may well occur in future contingency operations. But the injury rate for such jumps cannot be so high that it discourages training or limits operational use of this mode of force projection. A systems approach to the problem should assess all the contributing factors and work toward a comprehensive solution, which may have procedural and personnel components as well as a technology component.

More generally, a systems approach is appropriate when applying any of the softer sciences to the soldier's well-being and performance. In addition to using new learning techniques for training, the behavioral sciences are now developing ways to enhance the soldier's ability to deal with mission stress, fatigue, and environmental extremes. The STAR Committee recommends that a major systems effort be undertaken to determine and pursue those technologies within the psychological and medical fields that appear most likely to enhance the performance of the individual soldier.

Combat Power and Mobility

The technology management issues for this function fall naturally into the same three categories used in Chapter 2: long-range mobility, battlefield mobility, and lethal systems.

Long-Range Mobility

Two complementary aspects of long-range mobility deserve equal attention: air transport of immediately deployable forces and sea transport of reinforcing heavy forces.

The second aspect has received less attention in this report not because it is of less concern but because the report's timing precluded thorough study of the lessons from Desert Storm. The STAR Committee anticipates that there is much to learn about the movement of heavy forces from their bases to port, loading and unloading for marine transport, and deployment at a distant site of contingency operation. The Committee also expects that the Army is already studying these lessons and will continue to do so for some time. By way of encouragement, then, the Committee wishes simply to repeat a truism: getting heavy forces to the battle in a timely manner will be as important to ultimate success as getting the immediately deployable forces in place quickly.

Chapter 6 suggests a force structure transition toward a much larger echelon of air-transportable forces that would have enhanced capability to defend against opposing heavy forces. The following discussion refers to these proposed future "immediately deployable" forces rather than the current force structure of airborne and air-mobile units.

With respect to technology management in support of immediately deployable forces, the STAR Committee has several substantive recommendations. The transport problem will only grow as the Army becomes largely based in the continental United States, but the funding realities portend that fielding of a new long-range transport is unlikely. So the Army will need to focus on how best to use the long-range transport already available. Traditionally, advanced systems design has not treated the issue of how to get a system to the battle as a primary constraint; the Committee suggests that it must now become a primary constraint on design.

The category of available long-range transport has two elements:

• *Military transport systems.* The transport capacities of military aircraft available to the Army (currently the C5 and C141, the C17 in the future) should be assumed as design constraints on systems intended to accompany the immediately deployable forces of the future.

• *Civil Reserve Air Fleet (CRAF).* The STAR Committee suggests that CRAF is a resource the Army can exploit more fully in the future. To do so, the Army should go beyond passively "making do" with whatever capacity comes out of current or future CRAF arrangements. The Army should work actively to influence CRAF capabilities. Such influence can be exerted in two ways: (1) by persuasion of the parties involved (i.e., the commercial cargo carriers) and (2) by seeking legislative inducements that favor capabilities the Army will need.

The CRAF as an exploitable resource becomes especially important if the Army chooses to pursue the suggestion in Chapter 6 that immediately deployable forces include everything movable by air, regardless of its nominal organization and basing structure. In this case the military's own transport capability will certainly be inadequate; the Army will have to look to CRAF to make up the difference.

Battlefield Mobility

The Army's current Soldier as a System initiative is addressing the issue of transporting the dismounted soldier's load. The section above on Integrated Support for the Soldier notes the Committee's belief that some mechanical means of transporting heavy loads over difficult terrain will be required. The discussion here of battlefield mobility assumes that adequate attention will be paid to mobility for the dismounted soldier.

For the kinds of terrains in which the STAR Committee anticipates future contingencies, some form of heavy-lift, rotary wing vehicle will be needed. If the Army agrees that the need exists, a technology base and demonstrator program will be required because neither a program structure nor a technology base exists now for such a system. The engine size requirements differ from those for other helicopter uses. There is also the potential for robotics implementations (see Heavy-Lift UAV discussion below). Of these two concerns, engine size and related design and development is crucial. A cooperative program with the Air Force could serve well here. Either an entire engine appropriate for heavy lift or parts of one could come out of the Air Force's high-performance engine program.

With respect to ground vehicles, a key issue for technology management is to determine precisely what capabilities will be required of future armored vehicles to engage in the anticipated range of contingency operations. Continuing evolutionary improvements to existing systems, notably the M-1 tank, will provide the time needed to evaluate alternatives. It is not yet clear to the STAR Committee whether the *long-term* direction should continue evolution from the current generation or whether a radical departure will best meet future challenges. Presumably, a worthy radical alternative would offer a significant technological leap ahead. Before the Army makes a commitment that abandons further evolution from the present highly successful designs, the potential of candidate alternatives should be not only explored but also demonstrated.

Within this context the STAR Committee supports the conclusion of two STAR panels (the Mobility Systems Panel and the Power and

Propulsion Technology Group) that electric-drive systems, powered by advanced engines, offer technological gains that merit Army appraisal. A demonstrator program for these alternatives would allow their promise to be tested. However, the issue of sizing the systems appropriately for expected operations should be addressed and decided during the design phase, before limited resources are consigned to a demonstration effort.

Lethal Systems

Among the successes of Desert Storm were Army weapons capable of acquiring targets and destroying them, under battle conditions, before they were targeted themselves. Our systems were more accurate and more lethal at greater range. These advantages need to be maintained in future lethal systems, for they hold the key to winning without sustaining high casualties. Unfortunately, the systems that can maintain these advantages all involve major changes with significant costs. While successful systems will be crucial, it is infeasible to develop and field numerous new ones. The question therefore becomes: How does the Army get enough new advanced systems without having to pay the full development cost of each? The Star Committee suggests an "investment policy" that combines two complementary approaches:

• Look for the lowest-cost application or adaptation of Army-usable advanced systems that are already under committed development by other services. Possibilities here include the Air Force/Navy antiradiation missile program (AGM-88C as a successor to HARM), the AIM-9 series of advances in the Sidewinder family of close air-to-air missiles, and the AMRAAM (advanced medium-range air-to-air missile) successor to Sparrow. As an example of adaptation, the AIM-9 or AMRAAM might be equipped with a new booster, to make them serviceable in an Army ground-to-air role. By adopting or adapting systems developed under other budgets, the Army can direct more of its limited resources toward a few special-purpose, Army-unique systems.

• For those systems to be Army developed, choose from among the various alternatives by experimental testing of their comparative advantages. Not all the potentially worthy prospects can be pursued into expensive development phases, so experimental testing should be incorporated in the early selection-decision phases.

This proposed investment policy for advanced weapon systems has an added advantage. It fits well with a coherent, rational consolidation of the technology base proposed by OSD and supported by all the services. The most desirable (because it is least harmful) outcome of this consolidation—which must inevitably occur one way or another—is for the Army to use its limited resources to support just the infrastructure required by its unique needs, while borrowing much from the infrastructure to which its sister services have made similar commitments.

One last lethal systems area in which technology insertion will be of special importance to the Army is in mine and countermine operations. In the opinion of the STAR Committee—supported by the experience of Desert Shield and Desert Storm—potential enemies can be expected to pursue advanced mining technology relentlessly. This area is potentially so effective yet relatively inexpensive for an opponent that it is bound to receive attention. The Army should therefore have an equally vigorous program in countermine technology. The options are wide ranging and include both active and passive measures. Test and evaluation will be needed before deciding which directions to pursue. Given the potential implications of mine and countermine technology, the Army should consider an organizational elevation of work in this area.

Air and Ballistic Missile Defense

Currently the United States has only a limited capability to defend its deployed forces against even the relatively primitive ballistic missiles possessed by third world nations. Similarly, U.S. forces appear inadequately defended against hostile aircraft and cruise missiles possessing the next generation of LO technology. As Desert Storm showed, although air or missile threats are not significant against our mobile combat forces, they can be effective against troop concentrations and facilities in rear support areas.

New systems to provide the needed air and ballistic missile defenses will require a closely managed focus on the enabling technologies. However, a multiservice framework must first be established to avoid wasteful duplication of effort and disparate, ineffectual results. The STAR Committee recommends that the Army take the lead in initiating the integration of technology programs throughout the services for air and ballistic missile defense.

An important avenue for this integration effort is the SDIO (Strategic Defense Initiative Organization), particularly in light of its recent shift in focus from defense in a massive nuclear exchange to issues of tactical air defense. While SDIO continues in its present form, it represents an independent budget line for R&D in this key area. The

Army should move to exercise overt leadership within SDIO planning and programming activities, because Army forces will be, in the main, those to be protected from tactical ballistic missiles.

Combat Services Support

Health and Medical Support

The STAR Committee supports the following recommendations made by the Health and Medical Systems Panel:

* *Trauma treatment.* Develop one or more centers for research and training on the treatment of advanced trauma and care of trauma patients. The centers should be a cooperative effort of military and civilian authorities to capture the synergy of treating trauma patients from peacetime, civilian conditions as well as combat-related injuries.
* *Disease and injury prevention.* Promote R&D in biomedical sciences on the physiology of physical fitness; in pharmacology and biotechnology on development of new vaccines and antimicrobial drugs; and in psychobiology and neuroscience on cognitive abilities, motivation, and mental health.
* *Diagnostic molecules.* Promote R&D in biotechnology for (1) early detection, identification, and countermeasures to prevent or neutralize the adverse effects of chemical and biological warfare agents and (2) reduction of the health risks associated with environmental hazards.
* *Combat casualty treatment.* Promote R&D on protective and diagnostic/therapeutic aspects of integrated soldier support systems; on field medical systems that emphasize mobility and far-forward resuscitation; and on development of new prostheses and replacements for skin, blood, nerve, and bone tissues.
* *Medical information technology.* Maintain a technology watch for new medical developments and new technologies with particular relevance to the Army's medical needs; promote the use of computers for medical data management, medical modeling, displaying information, and medical research.
* *Infrastructure for military medical research.* Strengthen the Army medical R&D infrastructure to ensure that excellent medical research personnel are recruited and retained. Use collaborative programs with universities to accelerate research on militarily relevant aspects of infectious disease, neurobehavioral science, and molecular biology.

Simulation Technology

A valuable addition to the Army's existing and planned capabilities in simulation systems would be a larger facility for modeling ground combat with a high degree of realism. This facility should be available to both the R&D and operational communities within the Army. It could be used to evaluate new tactics and to explore the application and utility of new technological opportunities.

As noted in Chapter 2, this kind of simulation on a massive scale (in terms of the computational resources required) is a unique technological capability of the United States. It should be exploited as a military advantage as well. Given that contingency operations may require U.S. forces to confront an opponent on the opponent's home territory, the capability to simulate the terrain, vegetation, weather conditions, order of battle, and potential opposing tactics could compensate substantially for the home advantage enjoyed by the other side.

THE ARMY'S R&D INFRASTRUCTURE

The changing world situation and domestic environment will demand continued attention to the roles and missions of the Armed Forces in technology development. Recently, the Army has moved aggressively to restructure its in-house R&D to be more effective and productive in areas of advanced technology. The STAR Committee applauds the bold steps recently taken by the Army in its Lab 21 initiatives. However, the Army should not unnecessarily risk fracturing those areas where it currently has the greatest expertise by an over-rapid physical consolidation of facilities.

The Committee also commends the Army's technical management for its efforts to lead the services and the OSD toward a more effective focus for overall DOD research and advanced technology development. Related to this broader view of the technology base is the point that the joint nature of contingency operations implies joint development, particularly for C^3I and information distribution activities that have no obvious initiating agency. The Committee encourages the Army R&D community to continue to seek opportunities for leadership and support of joint development in these and similar areas.

This progressive R&D management style will need to continue unabated as new requirements and new programs evolve within the Army and the DOD. The report of the STAR TMDP Subcommittee contains several sections that address the Army laboratories and

R&D centers on broad issues. The STAR Committee recommends seven measures that bear directly on Army R&D infrastructure:

- Shift, over time, from centers that focus narrowly on individual combat arms to each center having a broader *capability* orientation.
- Ensure adequate organizational support for Army *basic research*.
- Improve the *work environment* in Army laboratories in ways that demonstrate to the Army's scientists and engineers that their work is highly valued.
- Make the most of limited funds for in-house R&D by promoting *exchange of information* with industry.
- *Attract talented technologists* early in their careers and provide progressive career advancement programs to retain them.
- Where possible, use *rapid austere prototyping* and related techniques in the design and development of both platforms and subsystems.
- Maintain a *worldwide technology watch* for advances in areas of science and technology with implications for Army capabilities and for potential enemy capabilities that will have to be countered.

Capability Centers

At present the Army R&D community is in the midst of a major consolidation program stemming from the realization that, as technological complexity increases, it will become more difficult to sustain a critical mass of competence in its diverse R&D structure. The Army 21 proposal now being implemented both consolidates its advanced technology programs (6.1 and 6.2 development phases) in a central "flagship lab" and more effectively combines the remaining product laboratories.

The STAR Committee endorses the general idea driving this ongoing Army R&D reorganization. The Committee also endorses the Army's initiative in Project Reliance, the multiservice commitment to heavy interdependence among the services in critical technologies and development capabilities. However, the STAR Committee believes that in the long run (after the current reorganization has been accommodated), the Army should consider going even further in bringing together its technologists and technological experimentation. We believe that the architectures of new Army systems will become increasingly interactive, just as the individual combat arms elements within the Army will also become increasingly interactive.

Even after the current reorganization, Army development laboratories will still be organized around elements of individual combat arms. The STAR Committee believes that a more effective means for incorporating advanced technologies into future Army systems will better serve the ever-increasing complexity and interactive nature of these systems. The Committee's suggested alternative is to support a small number of centers, each organized around a broadly conceived capability. The Committee recommends that the Army's long-range planning seriously consider this next step in integrating its technical base.

For example, perhaps five years from now, each of five newly defined centers might be devoted to one of the following broader capabilities: C^3I, missile systems, autonomous systems, human resources, and simulation. Under this approach, each of these super-centers would be responsible for its capability area throughout Army applications. The center dedicated to C^3I, for instance, would be responsible for steady improvement of the Army's C^3I/RISTA systems, including simulation and exercising, development of detailed plans, developing or buying needed equipment, maintaining surveillance over related technologies, and support of those technologies that cannot be otherwise obtained. Over the long term, the current system of Army laboratories could be integrated into these centers.

This approach, which parallels the Air Force's laboratory system, should provide more effective development of the complicated technologies now on the horizon. With all the new technologies envisioned by the STAR working groups, the Army will find it more difficult to acquire and keep a critical mass of technologists, together with the expensive support structure necessary for progress in these technologies. Some means of concentrating people and resources, such as this capability approach, seems inevitable.

Support for In-house Basic Research

Without strong support for basic research, the foundation for developing future technologies will be weakened. Before the initiation of the present Army 21 reorganization of advanced technology activities, the STAR Science and Technology Subcommittee reviewed the then-current Army organization. Suggestions for a reorganization similar in leadership structure to the Navy's Office of Naval Research were prepared for inclusion in the STAR Technology Forecast Assessment for Basic Sciences. This STAR review also considered the possibility of an Army flagship laboratory like the Naval Research Laboratory.

The Army 21 reorganization clearly aims at objectives similar to those expressed in the STAR review, although the means of accomplishing them appear to differ in detail from the model conceived earlier by the STAR group. The STAR Committee has not had the opportunity to study in detail the Army 21 mechanism for basic research. Nevertheless, it commends Army 21 in general as a strong move toward ensuring a major Army research capability. The Committee supports the concept of an integrated, flagship laboratory but cautions that the existing laboratory structure is fragile; care must be taken in the timing and method of formulation of this flagship laboratory.

From a wider perspective, the STAR Committee believes it is also necessary to modify the Army's current requirements process, in part so that all the Army's research managers will have clear direction on areas where the future military needs will be greatest. This and other issues related to the requirements process are discussed later in this chapter.

Improved In-house Laboratory Environment

The STAR TMDP Subcommittee reported that it had detected a significant increase in the Army R&D community's general sense of dissatisfaction with the work environment. Although this malaise is not entirely the Army's doing, the Army will nonetheless bear the brunt of its effects. If the dissatisfaction truly exists, it bodes poorly for the future of the in-house technical work force, just at the time when the Army will experience its greatest dependence on technological progress.

The widely discussed causes of this dissatisfaction include constraints in contracting procedures and work authorizations, unusually large funding fluctuations, numerous outside reviews, delays in equipment availability, inspections and audits, personnel ceilings, and salary caps. The STAR Committee urges the Army to continue its efforts to assess these and other possible causes of this dissatisfaction and address them in a way that demonstrates to its scientists and engineers that the Army values their work.

The creation of an environment of freedom and technical challenge within its laboratories and centers may be the single most effective antidote to this dissatisfaction. The Army has already done much of what it can do on its own to improve its technology work environment. The STAR Committee suggests, however, that the Army may need to lead a multiservice and DOD effort to convince the Congress of the importance of committing to a broad program of improvement. The Committee also encourages the Army to look within itself for examples of actions that have worked to relieve the frustrations of the current government work environment and to consider the experiences of the laboratories within other services and within the national laboratory structure.

Exchanging R&D Information with Industry

Faced with the prospect of tightened defense budgets, the Army will continue to seek the most effective uses for the R&D resources available to it. To avoid duplication of effort and ensure timely application of new basic research, the Army's in-house R&D programs must be carefully coordinated with similar programs conducted by its sister services, the national laboratories, other federal agencies, U.S. industry, and even the R&D establishments of our allies.

The Army is to be commended for its leadership in programs like Project Reliance. However, emotions are likely to run high when preservation of long-established capability is at stake; it will not be easy to consolidate in a way that may seem obviously correct from an abstract conceptual view. Still, the Committee hopes that the Army will persevere toward full implementation of Project Reliance, because preservation of a deep and well-equipped technology base will require substantial further focus of resources within the government community.

In addition, the Committee believes that more attention should be given to achieving the best long-term use of the limited discretionary resources of the defense industry. Much of the relevant industrial R&D is conducted by defense contractors under the Independent Research and Development (IR&D) Program. While the Army receives extensive information from the IR&D participants via these IR&D reviews, there should be more emphasis on an Army effort to provide this defense industry base with a greater level of detail on its own in-house R&D programs. The pilot programs to achieve this interchange, which were recently initiated, should be expanded, and the pressure for mutual government-industry sharing should be maintained.

Also, cooperative programs between government and private industry, like those being pursued at the Electronic Technology and Devices Laboratories, seem especially promising. The STAR Committee believes that these programs should become the model for a similar but much broader effort throughout the Army technology community and, for that matter, within the whole of DOD.

The Army has consistently been a leader in advocating legislative reforms to remove the legal obstacles to closer cooperation between industry and government. Still, an even stronger emphasis on such reforms seems necessary. We hope that the Army continues to spearhead that interest.

Attracting and Retaining Technological Talent

As the Army incorporates more and more advanced technology into its war-fighting capabilities, officers and enlisted personnel who understand these technologies will become increasingly valuable. The Army will also have to contend with demographic trends that forecast a smaller pool of well-trained young people to recruit. Furthermore, other studies project an increasing demand from all sectors of the economy for scientists and engineers at all degree levels. Thus, the Army will be competing with the civilian sector in attracting and retaining qualified engineers and scientists in its civilian and military ranks.

A successful Army R&D program will depend on attracting technical professionals even before they receive their advanced degrees. And for all its R&D personnel, the Army must offer innovative incentives to retain those with the most valuable skills. For enlisted personnel, the Army will have little alternative but to accept the responsibility for developing technical skills through expanded training; it has already begun to do so, with considerable success.

The Army now has a particular opportunity to establish the kind of career education and assignment program necessary for it to cope with the technology forecasts described in Chapter 3 and in the STAR working group reports. The Mavroules Amendment to the 1991 Military Appropriations Bill can become the vehicle for such a program. The STAR Committee encourages the Army to pursue the opportunity this amendment provides.

In particular, STAR suggests that the Army acquaint itself with the results of the apparently very successful Laboratory Demonstration Program conducted at the Naval Weapons Station, China Lake, and at the Naval Ocean Systems Center, San Diego. In these decade-long "demonstrations," remarkable improvements in both the quality of scientific talent recruited and retention of the best of that talent have been achieved. These centers have clearly retained reputations for producing relevant military technology of the highest quality.

Rapid Austere Prototyping

Rapid prototyping is the development, on a compressed time scale, of preliminary versions of the new components in an advanced system design. The prototype should include all of the unproven (hence risky) elements of the design. However, it needs to include only as many of the low-risk elements as are essential to prove the new concepts. For the latter reason, it is also called *austere prototyping*. Often, however, military prototyping programs aim not at an austere prototype suitable for testing risky concepts early in the design phase but at something closer to a preproduction version of the system.

Properly used, rapid austere prototyping can reduce the time between system concept and production by proving design concepts and pinpointing flaws in need of redesign early in the development cycle. It also lets the prospective user see what is possible while reducing or delineating the development risks. User input based on exercising a prototype is far more valid than requirements definition based on experience with old technology. In an era of explosive technology growth, it can assist the Army in fielding new technology while it is novel enough to give a distinct tactical advantage.

The Army's Technology Base Master Plan already incorporates significant opportunities for rapid prototyping methodology in its specific technology demonstrations and the Advanced Technology Transition Demonstrations (ATTDs). A specific technology demonstration is usually conducted in a laboratory environment during the 6.2 to 6.3A phases of development. It is used to provide information that will reduce uncertainties and engineering cost. The ATTD, which is conducted in an operational rather than a laboratory environment, is intended to provide an integrated proof-of-principle demonstration at the 6.3A phase, so that near-term system development can satisfy specific operational requirements.

For either type of demonstration, the major thrust of rapid prototyping methodology is lost if the test results, negative and positive, cannot feed back into redesign and even concept revision. In addition, the rapid prototyping approach needs to be diffused through all levels from components and subassemblies to systems. Ideally, the dozen or so current ATTDs would each represent a final, large-scale prototype test following on the lessons learned during multiple lower-level prototyping events, perhaps along the lines of the current specific technology demonstrations.

Aside from these reservations, the STAR Committee endorses the attempt being made through the ATTD program and other test and demonstration procedures to define the objectives of the Army's prototyping programs. It is not enough to carry out a technology demonstration program if its fruits do not arrive in the field in time to assure the technological superiority of U.S. forces in combat. From the STAR Committee's perspective, a prerequisite of any prototyping program must be the *preservation of continuity in the technological advantage* of U.S. Army forces at any time those forces may be asked to engage in combat.

The STAR Committee encourages the Army to think through how best to preserve continuously the technological supremacy it now enjoys. Global emergencies may well demand action against sophisticated and able enemies faster than technology can be fielded by any program that does not begin until the need arises.

Worldwide Technology Watch

The advances in technology occurring worldwide will be available to our potential adversaries as well as to U.S. defense forces. In this changing world, the Army will need technical and management preeminence to maintain tactical superiority. Achieving this preeminence in a period of budget pressure is a considerable challenge.

At the least, the Army will have to maintain a worldwide technology watch over advances in various areas of science and technology. This will require an understanding and sensitivity to the potential applicability of technology at all levels in the Army. There should probably be specific responsibility for this function designated within the Army technical community. The STAR Committee suggests that the Army consider how to implement this military technology watch and then commit appropriate personnel and funds.

TECHNOLOGY MANAGEMENT AND
THE ARMY'S REQUIREMENTS PROCESS

Late in August 1990, a special panel composed of members from the larger STAR panel met to consider the Army's requirements process as it applies to advanced technology utilization and force modernization. Despite the diversity of the participants' experience, they were able to achieve considerable consensus. The STAR Committee has adopted portions of the panel's analysis and conclusions and presents them in abbreviated form in the first four subsections below. The last subsection ties this assessment of the requirements process to the technology management strategies presented earlier in this chapter.

Implications of the New Environment
for the Requirements Process

As the Army progresses into the last years of the twentieth century, it finds itself subject to external circumstances that inevitably will strongly influence its force structure and the equipment it procures. This portion of the STAR main report examines whether the current Army requirements process can deal efficiently with these new external environments.

The current requirements process has evolved over several decades to set priorities for the Army's response to a scenario of Soviet confrontation that changed only slowly. Also, this requirements process evolved while resource levels were reasonably stable and while the Army expected to support combat operations primarily from its own resources.

As the current system evolved within this reasonably stable fiscal and threat environment, efficiency was achieved by parceling out the work of developing detailed requirements to the individual combat arms centers. The detailed knowledge and enthusiasm of the individuals at these centers was thereby fully utilized. The participants shared a fairly clearly understood, overall concept of operations, and this concept changed infrequently. Because the top-down policy constraints remained so stable, the Army Concept-Based Requirements System (CBRS) became, in appearance and substance, a bottom-up requirements system.

However, the external environments that now weigh heavily on future Army acquisition decisions are far less stable than previously. The STAR Committee finds the principal destabilizing factors to include the following:

• severe overall DOD budgetary limitations, leading to severe force structure reductions within the Army (shared also by its sister services) and significantly reduced Army acquisition budgets;

• a rapidly evolving and highly uncertain set of future threat scenarios, particularly when compared with the scenario of mid-European Soviet confrontation from prior decades; and

• the likelihood of far more intense, joint (multiservice) contingency operations than were previously required by mid-European scenarios.

The STAR Committee concludes that these new circumstances would probably stress the present Army requirements process in three ways discussed below.

1. *For some considerable time into the future, a more top-down requirements process will be needed.* In the severely limited fiscal environment postulated for the next decades, a higher degree of selectivity in approaches to be implemented will be required than before. The design of any one combat system will be more dependent than formerly on the characteristics of other systems with which it interacts. This ap-

plies not just to intersystem relations within the Army's domain but also to the fit of Army systems with those of other services and with the overall OSD architecture. Definition of elements of the future U.S. military force structure will require more active and continuing top-down guidance than has been the case.

From the Army's standpoint, this factor is compounded by the clear implication of substantially greater joint service interactions and interdependence. As future scenarios unfold, greater land-sea-air interfaces can be expected because of contingency geography, extended battle ranges of both our own and our adversaries' weapons, and the concentrated nature of our expected lodgements. As the three services become a more integrated set of combat forces, the weapon systems that each service projects for the future must become part of a common combat system architecture.

The analyses to support new directions in technology or systems must show how those changes fit with the overall architecture of OSD priorities. Otherwise the Army will continue to lose out in the allocation of resources.

2. *A greater requirements emphasis on cost/performance balance will be needed, both at the beginning of a program and through its lifetime.* In an environment of limited procurement, it becomes crucial to strike a balance between the capability required and the cost of that capability. During the last few years, all the services have been encouraged to seek optimum performance, knowing that inventories would eventually be built out to sufficient size. That assurance of eventual inventory build-out can no longer be taken for granted. Further, even if inventories can eventually be filled, the time frame of build-out may well be so extended that the service cannot wait for the capability.

The STAR Committee perceives a need for a requirements process with substantially more iterations than at present for balancing the military's needs against the cost of meeting them. The balance will need to be reconsidered both at the onset of each program and at intervals throughout the development phase of the program, while the state of the technology is still not demonstrated.

3. *More exploration of feasible alternatives should be done before a requirement is specified.* As budget pressures extend the time between fielding of model changes, more frequent opportunities will arise to explore by experimental, prototype demonstrations the real operational advantages of capabilities that previously were only imputed by simulation or computation.

The STAR Committee foresees a greater opportunity in the future requirements process for feasibility demonstrations oriented toward a generally acknowledged need, before convergence upon a formal requirement. The ATTD program (discussed above under Rapid Austere Prototyping) is an excellent start in this direction. The requirements process should expand on this start by using prototype testing to better evaluate what is needed and to take advantage of the extended time necessitated by longer design lifetimes. For this more iterated technology/capability process, the STAR Committee envisions a far tighter cooperation between the Training and Doctrine Command (TRADOC), representing the users, and the development community.

Changing the Requirements Process

The STAR Committee recommends six changes to the process by which requirements are generated and incorporated into the Army's program.

- *Keep the CBRS; alter the process.* The essential intent of the CBRS should be retained; the implementation must be radically altered. The next five recommendations pertain to specific alterations.

The Army may already be initiating some of these changes. In December 1990 TRADOC and the Army staff began a reassessment of the CBRS to make it more relevant in generating future Army requirements. As the STAR study was drawing to a close, this internal reassessment was just beginning; it was too early for the STAR Committee to determine how this initiative would affect the technology management problems described above.

- *Open up the front end.* The "concept" input to the requirements process should be opened up to technology exploration and to concepts built on advanced systems concepts and likely threat scenarios. The input to the process should also allow for broadly defined capability issues, such as force projection, force employment, and sustaining deployed force. Advanced systems concepts could aid in capturing these broad issues for consideration.

- *Ease up on Phase 1 specificity.* The current approach to delineating qualitative requirements, Required Operational Capability, presumes too much specificity too early. It should be replaced with something closer to the "materiel need" approach used in the early 1970s. The latter identified "must haves" and "wants" early in the requirements process, but it deferred final selection until data gathered during development could be factored into the decision process.

- *Winnow as you go.* The present understanding of accepting a concept-based requirement into the program is that anything put into

Phase 1 research is destined for eventual Phase 4 development. To encourage innovation, it is better to let Phase 1 be accessible to more players. Instead, increasingly stringent winnowing decisions should occur as part of the move to each subsequent phase. Thus, many Phase 1 research concepts will never move forward. Some Phase 2, and even Phase 3, systems will suffer similar fates.

• *Test, evaluate, and redesign.* Testing and evaluation are now often used to justify a program's legitimacy to the Army or Congress. They also become captive to the need to check compliance with contract specifications. The roles of test and evaluation need to be rethought in terms of subjecting systems to field conditions, learning from both the successes and failures during testing, and applying test results that capture design flaws in need of redesign. As just one example, the methodology called rapid prototyping, which was discussed earlier in this chapter, is one approach to pulling aspects of test and evaluation forward into the design process itself.

• *Provide a vision from the top.* If the concepts going into the CBRS are opened up to technology exploration and the standard of specific Required Operational Capability is relaxed, then control over program building must be exerted from another quarter, preferably from the top down. But heavy-handed management from the top (micro-management) can be as disastrous for innovative technology as narrowly conceived requirements definition from the bottom. By communicating a strategic vision from the top down, technology managers can guide the CBRS process while leaving individual "concept" origination open to an array of participants. In addition, the top of the organization must ensure that the rationale for each part of the program has been clearly linked to the defense policy architecture and priorities of OSD.

Organizational Realignments

The STAR Committee suggests three areas in which the Army organization will need realignments, if the process changes recommended above are to revitalize the CBRS.

• *Reassign control over requirements.* The combat arms centers should no longer drive the process by controlling the definition of requirements. Neither should they be excluded from the process. Instead, they should be active participants whose input includes their views on mission and system requirements. But the process must be controlled from the top and must be open to other contributions as well.

• *Broaden the contributor base.* Opening up the front end of the CBRS to more contributors must be accompanied by "invitations to participate." It will be necessary to cultivate organizations inside and outside the Army that can provide the kinds of concept inputs the combat arms centers cannot. The invitation should not be totally unconstrained. The technology assessments, threat analyses, systems concepts, etc., that are contributed to the CBRS must have clear links to the strategic vision. The presence of such a link would not guarantee adoption into the program or eventual advance beyond Phase 1; it does set a minimum requirement for legitimacy. Reconsideration of the linkage, in light of research results and changes in external factors, should be an integral part of the decision whether to promote a concept or system beyond Phase 1 and at each further step along the way to final fielding.

• *Assign a process manager.* The first organizational recommendation above leads immediately to the question of who, or what organization, should manage the Army's requirements process and program building. Another way to ask this question is: Who should be the keeper of the vision?

An Army Management Review that was issued in October 1989 and instituted during the subsequent year has resulted in a three-tier organization headed by the Army Acquisition Executive (AAE). The AAE is to be the integrator of all acquisition action. The organizations reporting to the AAE are intended to support not only systems acquisition but also technology assessment and development. This recent realignment may well decide who manages the process. The STAR Committee could not, however, assess the effects of the new organizational structure on technology development.

The Requirements Process and Technology Management Strategies

To conclude its assessment of the requirements process, the STAR Committee offers a few final reflections on the relation it sees between strategic thinking about technology management and the preceding recommendations for the requirements process.

• The strategic focus presented at the beginning of this chapter, fleshed out with its focal values and function-specific focal interests, exemplifies the kind of strategic vision that could guide the CBRS when it has been opened to innovative technology concepts at its front end. The primary concern is that this strategic vision must include technological judgment; it must express what can be accomplished if the available technical knowledge is applied.

• The STAR Committee has suggested that the process of building a program out of the inputs to the CBRS should be more than requirements-driven, more than a distribution of the resource pie among competing internal interests. The practical content of this "more than" is an implementation strategy. A focused strategy can provide implementation guidelines for whatever organization is assigned the task of building the program.

The particular focal interests or implementation elements suggested by the STAR Committee are certainly not the only plausible content for a strategic vision and an implementation strategy. Some may prove worthy of adoption by the Army; others may not. Still, the Army needs a vision to guide a revitalized CBRS from the top. And it needs a concrete implementation strategy to counteract implementation by consensus.

6

Technology Implications for Force Structure and Strategy

INTRODUCTION

The STAR Committee was formed to execute the three charges specified by the Assistant Secretary of the Army (RDA) in requesting the study: identify the advanced technologies most likely to be important to ground warfare in the twenty-first century; suggest technology strategies for the Army to consider in developing their full potential; *and project, where possible, the implications of these technologies on force structure and strategy.*

There is ample precedent for the third charge in this request. Throughout history advancing technology has profoundly affected the structure of military forces and the conduct of war. The STAR Committee agrees that it is appropriate to consider not only the evolution of capability through technology but also the influence of new capabilities on future strategies for their use and on force structure requirements. However, the forecasting of future strategy and force structure consequences is at best an uncertain art.

Background

From past examples it appears that full evolution of strategy and tactics in response to capabilities enabled by new technology has sometimes taken a long time, often as long as several decades. Frequently, full adaptation to these new capabilities occurred only when the exigencies of combat forced exploitation of the new technologies.

Yet in World War II and again in the recent conflict in the Persian Gulf, the United States relied heavily on recently introduced weapon systems. It adapted prevailing strategy to anticipate successful use of its new technology-based capabilities. In fact, the ability to use these new capabilities both strategically and tactically gave U.S. forces the dominance they enjoyed in the Gulf war. So it is not clear to the STAR Committee that the traditional delay in adapting military practice to newly introduced capability need be as long as it has been in the past. Another lesson of this recent war is that demonstrating the full military significance of new technology may prove vital to future deterrence and, if necessary, to future warfighting.

In the past, lack of confidence in the military utility of new technology applications frequently delayed their introduction. Today, such uncertainties can be substantially ameliorated by highly realistic simulation programs and by scored field testing. Therefore, the STAR Committee believes that the Army's future strategies and force structure should be able to adapt much more quickly to technological opportunities. The delays in technology implementation may depend instead on the ability to bring these technologies quickly to the field.

Levels of Technological Impact

Many of the technologies and system applications reviewed by the STAR panels will require a time frame of a decade or two before their influence can be felt. The STAR Committee expects that major near-term changes to both military strategy and force structure are more likely to be forced by the profound changes now occurring in geopolitical and economic realities. Yet the Committee also believes that the near-term effects of these changes can be influenced substantially by prudent application of available and emerging technologies.

Basic U.S. strategies and force structure probably will not change markedly over the next decades, just as they have not changed markedly over the past decades. Yet the STAR Committee does expect the details of both to change in response to new adversaries, to budgets, and eventually to the new technologies of greatest import, once these are fielded. For these reasons, the Committee has chosen to respond to the request that initiated STAR by discussing the significance of future technologies in two sections: expected near-term changes and expected long-term changes.

The expectations for each time frame will first be treated separately. Then a common thread of conclusions will be presented at the end of this chapter.

NEAR-TERM IMPACTS ON FORCE STRUCTURE AND STRATEGY

The STAR Committee believes that factors like those outlined in Chapter 1—actors external to technological advances such as geopolitical changes and domestic economics—will be the dominant influences on force structure for a time horizon out to about 15 years (until about 2007).

However, during this near term, new applications of current technologies can have important second-order effects. In particular, these new applications may be able to ameliorate some of the negative consequences of the political and economic factors and smooth out the ongoing transition in force structure.

The STAR Committee concludes that the following nontechnological stimuli will have the greatest influence on U.S. force structure and strategy in the near term:

- the demands of new contingencies—the potential for sudden crises that involve diverse adversaries, resulting in rapidly implemented joint operations of U.S. forces;
- anticipation of enemy responses to the Persian Gulf—the responses of potential adversaries to the capabilities they see as responsible for the overwhelming U.S. victory in the Gulf; and
- The new political and economic situation—the combined effect of U.S. force reductions under the Conventional Forces in Europe Treaty, termination of basing rights elsewhere, domestic base closures, and continued budget pressure to reduce expenditures with delay, deferral, or cancellation of desired new Army capabilities.

Each of these three stimuli will probably result in modifications to both Army strategies and force structure. In fact, at the time of this report, all the services are examining how best to proceed in this new environment. Emerging technologies can support this ongoing Army response in the following ways.

The Demands of New Contingencies

These new contingencies are likely to differ from the scenarios of the past four decades in terms of more rapid evolution and relative unpredictability of who the adversary will be, where the confrontation may occur, and what presence the United States will have in the area prior to the time the contingency arises. The United States probably will continue its successful two-tier strategy of rapid response with immediately deployable sea, land, and air forces, followed by sea-lifted heavy forces for assault of any large and heavily armored opponent. Technology should, in the near future, be able to augment execution of this durable strategy in the following ways:

- better rapid contingency battle planning through advanced computer capabilities;
- better logistics support for rapid deployment, through advanced automated planning;
- better and faster training for characteristics of the contingency area, through digital terrain modeling and computer-aided instruction on the capabilities and attitudes of the opposing force;
- provisioning of greater combat power to initially deployed forces through advanced antiarmor capabilities: LOSAT (line-of-sight anti-tank), AAWS-M (advanced antitank weapon system—medium), terminally guided MLRS (multiple-launch rocket system), and so on;
- greater interservice dependence to solve time-phased initial deployment deficiencies (such as electronic warfare) through the development of joint procedures and training programs;
- better use of available C^3I (command, control, communication, and intelligence) information and IFFN (identification of friend, foe, or neutral), through better automated data fusion and application of software network control technologies; and
- improved concurrent joint battle operations through joint battle modeling, simulation, and training exercises.

Each of these seven technology supplements to current strategy and force structure are discussed in Chapters 2 and 5.

Anticipation of Enemy Responses to U.S. Successes in the Gulf

Potential adversaries throughout the world are surely considering how best to obviate the conditions that allowed so dominating a success for U.S. forces in the Persian Gulf war. In turn, the Army must try to anticipate and obviate these counterstrokes. To win a war against the United States in the immediate future may not be a realistic consideration. However, other alternatives are open to a potential opponent.

Perhaps the most straightforward way to deter the use of U.S. force is to vastly increase the probable casualty rate to U.S. forces, with the expectation that U.S. public opinion will not long support U.S. action in such a situation. The threat to use nuclear weapons on the battlefield—for example, by delivery from a tactical ballistic missile—could present a considerable problem to the United States and the Army. It would, in effect, reverse the roles with respect to the use of tactical nuclear weapons that the United States and the Soviet Union played in the past. Overwhelming conventional force was then held hostage to the threat of escalation to nuclear weapons as a last resort. Such a situation could occur in the future but with U.S. conventional forces playing the hostage.

How realistic such a nuclear scenario will be depends on political agreements unforeseen at this time and is well beyond the scope of this STAR study. However, even using conventional munitions to put U.S forces at greater risk prior to a war still seems a more attractive alternative for a potential adversary than attempting to win a war of direct confrontation with U.S. forces. From this perspective, the following threats of high casualties, made possible by technology that will soon be widely available throughout the world, seem potentially advantageous to an adversary:

- urban guerilla attacks on U.S. troop installations, analogous to the attack on the Marine barracks in Beirut, Lebanon, after which the attackers hide among the noncombatant population, so that U.S. forces will refrain from retaliation to avoid large numbers of noncombatant casualties;
- improved methods for use of chemical and biological warfare agents;
- low-flying cruise missiles to attack rear-echelon infrastructure;
- advanced, but available, tactical ballistic missiles capable of surmounting our current defenses;
- tanks with more recent technology than those used by Iraqi forces during the Persian Gulf war, as a means to avoid being outranged and outgunned;
- intense jamming of battlefield identification in hopes of causing excessive fratricide; and
- attacks on initially deploying U.S. forces before U.S. heavy forces can reinforce them.

The STAR Committee suggests the following near-term programs as representative of responses necessary to counter the reactions of potential adversaries to U.S. successes in the Persian Gulf war:

- Implement as a top priority the ensemble of programs constituting the Soldier-as-a-System initiative proposed by the current Army R&D Master Plan.
- Include passive and active measures to defend against guerilla-

style attacks in contingency wargaming and scenario analysis. Active measures include development of human intelligence assets, deception and misinformation activities, and other psychological operations directed against armed resistance fighters and their supporters. Weapons technology for noninjurious incapacitation will be needed to deal with opponents who use noncombatants or hostages as shields against retaliation with deadly force.

• Apply language training technology to reduce dependence on "friendly" foreign nationals for translation and, perhaps more importantly, for interpretation of an unfamiliar culture.

• Integrate real-time direct communication links between AWACS (airborne warning and control system) and Army low-altitude air defense elements to allow maximum use of Stinger, Chaparral, and Hawk for intercept of enemy cruise missiles. Consider adding rotary wing antiair protection of major logistic concentrations as a more effective way to protect against low-flying cruise missiles.

• Augment the lethality and engagement volume of Patriot to the greatest degree possible. Continue to support the SDIO (Strategic Defense Initiative Organization) program for theater air defense. Begin to shift the emphasis and positioning of Patriot force structure from air defense to anti-TBM (theater ballistic missile), and move toward greater dependence on Air Force aircraft for high-altitude air defense.

• Implement, where possible, greater stand-off range for existing U.S. direct-fire antiarmor systems (TOW, Hellfire, etc.). Introduce LOSAT, particularly to early-deployed forces.

• Focus technological effort on the "reduced difficulty" problem of unambiguous real-time IFFN by assuming continued U.S. air superiority and known (through the Global Positioning System) location of U.S. forces. This problem may be far more tractable than the more general IFFN problem dictated by the prior European scenarios, where air superiority was not assumed.

• Focus current technology on techniques for operationally acceptable bandwidth reduction of voice-actuated information to avoid circuit overload conditions similar to those experienced in the Gulf. Further, extend to the Army current commercial procedures and techniques for dealing with extreme peaking of circuit usage.

In conclusion, the STAR Committee suggests a near-term Army strategy that assumes a vigorous attempt by potential adversaries to deny the United States the low-casualty successes of the Persian Gulf war. This strategy focuses the Army scientific community on counteracting such attempts. The STAR Committee further expects that eventual force structure changes will have to be made to counter these anticipated enemy responses.

The New Political and Economic Situation

The reduced resources expected in the next decade will obviously have direct effects on the force structure of the Army. The size of these reductions may be extreme when compared with the array of forces potentially at odds with future U.S. national interests.

One method to accommodate some of the expected reductions in force structure will be to provide increased lethality to future ground forces. Another is to better use C^3I for improved use of ground forces remaining after downsizing. Both of these options have been discussed at length in preceding chapters of this report.

Another possibility with merit, despite its considerable organizational difficulties, would be for the services to consider a mutual ceding of functions that are now performed redundantly by several of them. The reductions in resources and the changes in the threat situation justify a close look at current allocations of missions and responsibilities to determine whether existing redundancies are still appropriate. Because U.S. forces are likely to be fighting opponents other than the Soviets, advantages will exist that could not be assumed under previous scenarios. As one example, U.S ground forces engaged in contingency warfare can expect to have overwhelming air superiority, whereas Warsaw Pact air power in the central European theater formerly outnumbered NATO (North Atlantic Treaty Organization) planes. This and other differences in circumstances are great enough to justify a fresh look at effective ways to apportion responsibilities. A candidate for such treatment is high-altitude air defense, as mentioned above. Given the expectation of U.S. air superiority, the Army might gain by shifting forces from this area to other combat arms.

Other candidates for consolidation are specific intelligence roles and electronic warfare assets. Pursuit of the current Project Reliance may well lead to these consolidation efficiencies, which in turn could lead to force structure strengthening.

Other possible means to mitigate the consequences of cuts in Army force structure include technology-based actions. Each action suggested below is aimed at better distribution of remaining forces. (This list is only partial and suggestive of the possibilities.)

• A shift in balance between combat and support forces may be possible through a shift in Army technological emphasis and procurement practices from maximum operational performance to substantial reduction in the requirement for repair and maintenance.

• A focused reduction in the uniformed technology training base could also result from this reduced need to repair and maintain equipment in the field, particularly if combined with far broader (and more economic) use of civilian "technical representatives" for first-level technological support to combat troops.

• Procedures for ordering from the field could be substantially automated, similar to the automated inventory management now widely used in the private sector.

LONG-TERM IMPACT OF TECHNOLOGY ON FORCE STRUCTURE AND STRATEGY

Although the immediate changes in the Army's force structure and strategy will be driven largely by the previously discussed external political and economic factors, the STAR Committee foresees technology exerting a far greater influence in the longer term. The Committee expects that technology will reinforce the trend toward a smaller but more capable and highly transportable force. It presents here six significant effects that advanced technology may well have on force structure and strategy during the last decade of the STAR 30-year forecast horizon. Some of these effects are continuations from factors that were reviewed above for the near term. Others will result from technologies that will first become accessible to the Army in the longer term.

The following long-term consequences of technology for force structure will be discussed:

• augmented information superiority;
• flexible, multiple-tier force structure for combat power and deployment;
• integrated defense against next-generation air threats;
• evolving role of rotary wing capability;
• support and maintenance allocations; and
• training scope and methodologies.

Augmented Information Superiority

Winning the information war in future combat contingencies will remain as vital as it has always been in the past. U.S. success in the Persian Gulf is a compelling example of the benefits of such an information "victory." The STAR Committee believes it should be a dominant strategy of the Army to win the future information war and to take the steps necessary to do so.

The Army's military strategy must continue to use superiority in information management to allow superior maneuver of its forces and the application of overwhelming force against the opponent. The Army must therefore commit to the force structure and architecture necessary to achieve information superiority. In future wars, however, the advantages of the United States in information technology may not be anywhere near as complete as was true in the Persian Gulf war. The Army needs to continue focusing on how to extend its information capabilities, both in anticipation of improved opposition capability and as a way to mitigate expected force structure reductions.

Space assets may well be available to our potential adversaries, either by their direct ownership or by arrangements with friendly noncombatants. In addition, the adoption by other nations of stealth techniques for air vehicles, which the STAR Committee anticipates will occur, may allow opposition air reconnaissance despite overall U.S. air supremacy. Application of low-observability techniques to rear support areas and assets can make such air reconnaissance more difficult and less rewarding for an enemy. Human intelligence gathering, deception and misinformation operations, and psychological operations can all be exploited in contingency situations if the requisite force elements for them are trained and available.

Even so, the new communication and sensor technologies forecast by the STAR panels offer the Army an unusually fine chance to extend its information lead rather than lose it. To do so, the Army will need to extend and broaden both the technology and the flow of information in two organizational directions. First, information capability and the information itself must move downward within the Army's own structure, so that even the smallest fighting units have a broad base of externally derived information. Second, information availability must extend upward to provide a greater participation in the future integration of all service and national intelligence sources.

For the downward, internal expansion of information availability within all elements of the Army, an architectural conformity must be imposed beyond what now exists. This architectural commitment will be reflected in the designs of all future new weapon systems and C³I programs.

For the upward expansion, each service, including the Army, will probably have to accept less-than-optimum performance of its intelligence information systems, when viewed solely in terms of service-specific mission requirements, in favor of improving the integration of all these systems. The STAR Committee believes that the combination of reduced resources and increased cost of new reconnaissance systems will force all of the major participants into more cooperative planning for the intelligence systems of the future. Because the Army will continue to be the principal battlefield user of real-time intelligence, the Committee recommends that it seek to lead (as it has done in Project Reliance) this future architectural integration of intelligence capabilities.

So far this discussion of information superiority has focused on electronic intelligence, as did the description of systems concepts in Chapter 2. Another side of the information war is human intelligence, or HUMINT, in which U.S. forces do not have the degree of overwhelming superiority they enjoy in the "high-tech" aspects of intelligence. In certain kinds of contingency operations—low-intensity or guerilla warfare, for example—this relative weakness in HUMINT could prove deadly. For example, the intelligence war fought in the Persian Gulf in 1991 was an electronic war, and an entire reinforced Marine *division* suffered 24 killed in action. By contrast, the earlier contingency operation in Lebanon was much more of a HUMINT intelligence war, and in that operation a single reinforced Marine *company* had 239 killed in action.

Certainly, many factors entered into these two disparate outcomes. But the contrast does underscore that HUMINT will remain important even in a "high-tech" Army. With respect to force structure, two points are worth noting:

• When U.S. forces are deployed to a foreign setting, specialists who speak the indigenous language(s) and understand the culture should accompany both combat and support units. U.S. forces should *not* rely solely on indigenous allies to provide all translation and interpretation.
• To provide this force component, the Army should investigate technology for more rapid acquisition of language skills and cultural training.

Flexible Multitier Force Structure for Combat Power and Deployment

Army discussions in progress are recasting the strategy and force structure necessary to respond to potential threats and to the budget constraints on overall size of the force structure. These discussions appear to be focusing on a multiple-tier level of forces, ranging from very light forces appropriate for Special Forces and "first-in" major contingency assignments, to "next-in," air-transportable medium forces, to the "later-in," heavier, sea transportable forces needed for assault of opposition heavy-armored forces.

Of these three tiers, the light and heavy forces will more closely resemble their present-day counterparts with respect to deployability, logistics support requirements, and relative scale of weapons systems. Of course, new technology will enhance their $C^3I/RISTA$, combat power and mobility, and air/missile defense capabilities far beyond their current counterparts, but their force structure characteristics will evolve naturally as the technology evolves.

For the middle tier, however, new conceptions of force structure must be forged, along with the technology and systems to support them. These medium forces will be rapidly deployable by air transport, but they must also be able to hold ground against heavy armor until heavy forces can be inserted. (Support from Air Force and Navy elements would, of course, be essential in this capability.) To build the middle tier, the offensive and defensive capabilities of current air-transportable forces will have to be substantially augmented. Another approach is to "lighten up" current armored or mechanized forces to the point that they meet the constraints on deployability and logistics. In practice, both approaches will probably be needed if sufficient strength in these medium forces is to be attained.

The basic principle underlying the medium force concept is simply the requirement to concentrate forces in space and time, applied in a context of rapid response to a range of potential ground warfare contingencies located far from bases in the continental United States. These medium forces will characterize the general-purpose Army of the future. Fortunately, many of the advanced technology opportunities forecast by the STAR panels can be applied to systems needed by this middle tier. Among the many examples are smart munitions for attacking hard targets; lighter, stronger, tougher materials designed for demanding applications; hybrid propulsion systems whose basic components can be used in many vehicle types and platforms; and robot vehicles for RISTA or "intelligent" missile and mine warfare. In short, the situation demands it, and the technology supports doing it.

Although it is convenient to think in terms of three distinct tiers of force structure, these tiers cannot become fixed in rigid organizational hierarchies. The potential variety of contingency operations, combined with constraints on total force size, requires the flexibility to allocate forces as needed for a particular contingency. In time, the technologies examined by the STAR study can support this flexibility. Among the potential applications, the following seem particularly important to the STAR Committee:

- the ability, through advanced computational techniques, to plan rapidly for flexible detachment or attachment of combat elements from one organization to another; in addition, through these same technologies, to plan and implement proper logistics support of these rearranged combat elements; and
- the ability for realistic training, through advanced remote simulation techniques, of the flexible combat structures just described.

In the preliminary STAR management panel discussions with retired senior military commanders, there was considerable insistence on this organizational flexibility. By enabling, in principle, training through remote simulation and planning for deployment and sustainment, advanced computer and display technology may well make possible a "mix and match" of forces to the task at hand that serves the Army's purposes better than the traditional permanent structure, which evolved when movement was less easy.

As an example, if conditions were extreme enough, the air-transportable combat power of normally sea-lifted divisions (MLRS, rotary wing aircraft, etc.) might in an emergency be transported by air and attached to the divisional structures of elements already on line. They could be reattached to their parent upon its later arrival in theater. In this way, a base of medium forces could be augmented with "extractions" from heavy forces.

Integrated Defense Against
Next-Generation Air Threats

The air threat to U.S. ground forces may well become increasingly diverse and lethal beyond 2010. By then it may no longer be possible to rely on the overwhelming air superiority achieved during the Persian Gulf war. In addition, advanced tactical ballistic missiles of considerable capability may well be available to any opponent with the resources to buy them.

The STAR Committee believes that the Army must develop a strategy, and eventually a force structure, to contend with these prospective capabilities of potential opponents. Since none of the current capabilities of the Army (or the Air Force for that matter), nor their immediate extensions, can be expected to cope with these prospective threats, this effort will not be trivial.

Not since 1864, when Sherman detached himself from his logistics base at Atlanta and marched to the sea, has a modern army been able to sustain itself for long without a large fixed base of operations. Unfortunately, for the contingency operations currently contemplated, these logistics lodgements may have to be especially concentrated. This concentration of logistics capability will be an important future vulnerability, unless a method can be found to contain the threat from stealthy air breathers and high-performance tactical ballistic missiles. Application of low-observable technology to support assets will be one necessary line of defense. Another will be active countermeasures to incapacitate or destroy attacking aircraft and missiles.

The STAR Committee expects that, within a decade, well-financed opponents will have procured cruise missiles and aircraft that use at least first-generation stealth techniques. By the latter part of the STAR time horizon, advanced forms of low-observability probably will have proliferated widely. In addition, within a decade a broader proliferation of advanced tactical ballistic missiles can be expected, perhaps including decoys or re-entry maneuverability, to make their engagement more difficult.

As a further force structure consideration, the diverging requirements for successful defense against both low-cross-section, air-breathing systems and long-range, high-speed ballistic systems probably will mean that a single system, with its accompanying crew, cannot continue to satisfy both requirements, as the basic Patriot has done. Two distinct systems, with their separate force structure requirements, probably will be required for success in both of these defense missions.

A technological solution to all these threats will require systems not currently available and a networked architecture, which is not yet implemented, for early detection, weapon assignment, and intercept. The extended battle zone for these kinds of engagements will also require internetting of capabilities resident in all three services as well as those in the national information community.

To date, there does not seem to be adequate attention within DOD to this severe stealthy-threat problem. The STAR Committee suggests (1) that the Army, as the service potentially most affected, lead the effort to define a program; (2) that the Army plan for the eventual implementation of a force structure to support this needed defensive capability; and (3) that low-observable technology be applied to concentrated support assets, as another means of decreasing their vulnerability to at least some modes of attack.

The SDIO and its deputate for Theater Missile Defense are pioneering a new range of system elements aimed at defense against ballistic missiles. As noted in the discussion of air and missile defense in Chapter 5, the focus of SDIO work has shifted from defense against a massive nuclear strike to tactical and theater air defense. The Army has now been assigned major elements of their program, especially those most applicable to future tactical air defense. The Army should encourage and support these SDIO programs, both with its best talent and in congressional testimony. Overall, however, the Army needs to think through its future focus in air defense and chart a course for its undertaking.

The Evolving Role of Army Rotary Wing Capability

In the view of the STAR Committee, rotary wing components of the Army force structure are likely in the far future to perform much less scouting but more heavy-lifting roles. Helicopters will almost surely continue to be used in some gunship roles and for inserting infantry and special operations forces into enemy rear areas.

The STAR Committee anticipates that in the far future the scouting role can be adequately performed by UAVs, which offer better survival against strengthened enemy air defenses without risking crews and expensive man-rated machines. The substantial obstacles to developing low-observable rotary wing vehicles, and their probably considerable cost, are further arguments for a long-term emphasis on UAV development and implementation of the force structure to support UAV operations. Because of both cost considerations and increasing enemy air defenses, the STAR Committee sees the proposed LH helicopter as perhaps the last generation of manned rotary wing scouts.

Augmented heavy-lift capability by rotary wing aircraft will be needed in many contingency areas where road and air base infrastructure may not be available. V/STOL (vertical/short takeoff and landing) substitutes for heavy-lift rotary wing systems appear feasible but expensive. In this sense, perhaps, there will be an eventual exchange of force structure assignments within the Army air community.

Support and Maintenance Requirements

For reasons of both cost and effectiveness, the Army of the future will probably radically downsize its force structure for logistics support in handling consumables and for repair and maintenance. Small numbers of cost-effective smart weapons will inevitably replace large quantities of dumb steel, as affordability techniques allow their procurement. Trends in civilian industry toward improved product durability and reliability will lead to lower repair and maintenance needs for Army platforms. This, too, will permit downsizing of the associated force structure element.

The STAR Committee also predicts that civilian contractors working as technical representatives will increasingly replace Army repair and maintenance personnel because of the substantial cost savings involved. The cost effectiveness of using contractors, whose productive working careers are several times longer than their uniformed counterparts, appears to the Committee to be eminently sensible for the expected period of reduced resources.

As a strategy, the Army should attempt to emulate the civilian world in its push to increase product reliability and durability and at the same time radically reduce repair and maintenance costs.

Training Scope and Methodologies

As previously noted, improved training will continue to be one of the best ways to improve the efficiency of deployed forces. The Army needs to be continually seeking methods to improve its training capability.

The STAR Committee has a suggestion concerning force structure in this area that may address several problems at once. As stateside forces increase, because of reduced overseas basing, and as cost pressures simultaneously force stateside base closures, there may be merit in diverting some returning force elements into experimental test units. These units could have a role—similar to that of the permanent Navy VX squadrons—in performing both operational and development evaluations. New tactical concepts and prototypes of new weapon systems could be evaluated by these units if they are based at or near existing Army development and test sites. The cost effectiveness of these tactical evaluations would be increased by interaction with progressively more sophisticated simulation technology.

The combat readiness of these detached units need not be appreciably reduced, for they can participate in remote, networked simulations and wargames. They can also participate physically with their parent organization in maneuvers.

Another major force structure implication of the Army's advanced technology for training is its potential role in reconstituting a full-strength fighting force should the need arise. Just as careful planning for use of commercial components and subsystems can provide important surge capacity in a sudden mobilization of U.S. military production capacity, so the steady attention to improved, automated methods of training reservists and new recruits can provide the surge capacity to increase total force structure.

Among the lessons to be learned (or reinforced) from the Gulf war are two that relate to the training of reserve forces:

- Both the training of reserve combat forces and the technology used in that training need improvement. Networked computer simulations, which allow units at remote locations to "train" and wargame with active forces, are one approach to be actively pursued. Software for a battle control language, described in Chapters 2 and 4, will add detail and realism to these training simulations.
- The existing emphasis on moving specialty areas to the reserves while retaining combat forces as active units should be strengthened. The National Guard and the Reserves are an ideal situation for specialties that depend on skills and understanding gained over time. Combat, on the other hand, places a premium on the straightforward physiological attributes, such as strength, sensory acumen, and physical endurance, possessed by younger soldiers in active units.

A continuing commitment to improve the training of total Army personnel, reserves and actives alike, and the technology used in that training, will not only improve existing Army forces but also create the capacity to "surge" production of well-trained soldiers in a wider emergency.

7

Conclusions and Recommendations

CONCLUSIONS

The STAR Study Committee draws the following 26 conclusions from the material presented in the body of this report and from the supporting STAR reports.

The Environment

1. The national interests of the United States continue to require a strong military force capable of being deployed anywhere in the world. A key part of this capability will be the Army's capacity to project dominating ground combat strength as a way to stabilize future regional unrest.

2. Military technological superiority, as demonstrated in Operation Desert Storm, will be required to ensure the dominating combat strength, minimal U.S. casualty rates, and avoidance of noncombatant casualties necessary for continued public support of U.S. force deployment in regional instabilities.

3. The explosive rate of technological progress observed in the last three decades can be expected to continue, if not increase, during the next three decades. Weapons of 20 years from now will have completely outmoded those of today, just as those of U.S. forces outmoded the older weapons of Iraqi forces in the Persian Gulf war.

4. Although most potential adversaries will not themselves possess the skills to develop and manufacture sophisticated weapons,

256

they may well possess the resources to purchase such weapons. To maintain the deterrence that comes from technological dominance, the Army must maintain a steady pace of technological improvement in its weapons systems.

5. Potential adversaries will study the conduct of the Persian Gulf war for lessons in how to counteract or neutralize the U.S. military capabilities demonstrated there. To deter U.S. intervention in a regional conflict, they are likely to seek means to threaten high U.S. casualty rates. The Army must be prepared to counter these measures.

6. To assure the continuing technological superiority of U.S. ground forces, the Army must maintain a strong technology base program. With the rapid progress in many diverse technologies, the Army's resources will be inadequate to fund all the significant opportunities. The Army must therefore draw widely on technology development programs within other services, the defense agencies, the national laboratories, and the private sector. In particular, the ability to reconstitute a full fighting force will require reliance on commercial production facilities.

High-Payoff Technologies and Systems

7. Significant opportunities exist to apply scientific and technological advances to military problems. With the proper focus of Army R&D resources, these opportunities can ensure continued technological superiority of U.S. forces. From a list of more than a hundred technologies with significance to Army applications, the STAR Science and Technology Subcommittee selected the following as most likely to produce important changes in ground warfare:

- multidomain smart-sensor technology;
- terahertz-device electronics;
- secure, wide-bandwidth communications technology;
- battle management software technology;
- solid state lasers and/or coherent diode laser arrays;
- electric-drive technology;
- genetically engineered and developed materials and molecules;
- material formulation techniques for "designer" materials; and
- methods and technology for integrated systems design.

8. From its review of the many advanced systems concepts used by the STAR systems panels to formulate battlefield capabilities and requirements, the STAR Committee selected the following systems as

having especially high potential as applications of advanced technology:

- robot vehicles (air or ground) for C^3I/RISTA;
- electronic systems architecture;
- brilliant munitions for attacking ground targets;
- lightweight indirect-fire weapons;
- theater air and missile defense; and
- simulation systems for R&D, analysis, and training.

9. Several systems characteristics were repeatedly cited by the STAR panels as benefits of systems for diverse functions. These traits were also advocated by participants in STAR symposia on future threats and Army requirements. The STAR Committee identified the following pervasive, beneficial characteristics of systems as *focal values* for the Army's technology program:

- affordability;
- reliability;
- deployability;
- joint operability (with the other military services);
- stealth and counterstealth (to reduce the vulnerability of U.S. combat and support systems);
- casualty reduction (among U.S. combat and support forces and among noncombatants); and
- support system cost reduction.

10. The Strategic Defense Initiative Organization (SDIO) has shifted from its original focus on strategic defense in a massive nuclear exchange to broader concerns with air defense capabilities in threat scenarios like those considered by STAR. Although the STAR technology assessments were largely completed before this shift occurred, the Army clearly needs to incorporate the new SDIO interest in tactical defense with its own planning for theater air and missile defenses.

Technology Management

11. Military systems incorporating advanced technology will be acquirable by potential adversaries. To be prepared to face adversaries armed with these systems, the Army needs more rapid fielding of critical capabilities based on new technology. The current Army programs for laboratory technology demonstrations and for ATTDs

(Advanced Technology Transition Demonstrations) under operational conditions are suitable vehicles for extensive use of rapid prototyping methods. The essential element is to feed early, austere prototype test results into design-redesign iterations before a systems concept is at the point of full-scale development.

12. Preservation of the front-end scientific and technical advantages provided by U.S. universities and industrial laboratories is essential to maintaining U.S. dominance in military hardware. The infrastructure dedicated to work of particular interest to the Army must not be permitted to deteriorate as declining resources diminish the support to these laboratories from the defense industry.

13. There is evidence of substantial disaffection with the Army among entrepreneurial small businesses and creative elements of larger industries. Increased Army use of this highly innovative part of the private sector appears crucial to rapid introduction of new technology into Army equipment. The Army's relations with this sector have been hurt by its rigid management. It has been unreceptive to concepts that do not fit its requirements as defined under the CBRS (Concept-Based Requirements System). It has insisted on ownership of intellectual property and unproductive competition. The motivations and attitudes of this community need to be considered and their cooperative support solicited. A greater emphasis on means of working together appears warranted.

14. Budget restrictions will limit the rate of introduction of future Army platforms. The design lifetimes of fielded equipment will therefore be extended. The Army will need techniques for timely introduction of new capability into its forces during the duty life of major platforms. This must be done in such a way that a potential enemy at no time can deny U.S. forces the technological advantage they now enjoy. In particular, systems must be *designed for change*, so that their subsystems and components can be upgraded during the extended lifetime of a system design.

15. A reduced Soviet threat and changing national budget policies are expected to lead to substantial U.S. force structure reductions. The current extent of forward basing, upon which the United States has relied heavily since World War II, is also likely to diminish. The methods through which the Army implements technology can ameliorate these difficulties while increasing performance. Specifically, technology must be applied toward enhanced combat power and mobility, plus rapid mobilization of forces (both manpower and supporting equipment). The objective of rapid long-range mobility must be given operational definition by the Army, then incorporated into

the early design phase of the system development cycle. For example, light systems suitable for air transport and sustainment can be designed to be more mobile tactically and more survivable yet have more firepower. As advances in technology provide the means to lighten heavy systems, the potential will increase to transport at least some of these systems by air in support of initially deployed forces in contingency operations.

16. Neither the Army nor its military contractors will have resources adequate for continuous retooling of all facets of the domestic military industrial base. In the modernization of processes and tooling, the Army's current specialized base of defense contractors could fall behind foreign competitors, who rely more heavily on industrial suppliers for their military products. To counter this trend, the Army can concentrate its diminished resources on those technology areas that have no private sector counterpart while depending to the fullest extent possible on commercial components and production facilities. The STAR Technology Forecast Assessments provide details on specific technology areas that are likely to be developed for commercial markets and those that will require Army support if they are to achieve their feasible growth.

17. The Army's equipment and systems requirements will remain so diverse that the anticipated levels of development and production funding will not support all of them. For this reason, as well as others cited above, interservice participation in major weapon systems development, as represented by Project Reliance, will grow in importance. Each service, including the Army, can rely on other services for some weapon systems while being the common provider for others. In this way the Army can release resources that would otherwise be tied to support of technology bases substantially paralleling those of other services.

18. Continued consolidation of the Army's internal technology infrastructure appears appropriate as the best way to maintain a critical mass of technologists in areas of Army-unique interest. Also, this concentration will free resources needed to procure the expensive yet necessary equipment required for advanced work in these technology areas.

19. Affordability of high-performance technology will be a crucial issue throughout the period to which the STAR reports apply. The use of technology to reduce the cost of systems that incorporate new technological capability appears both necessary and promising. Appropriate use of new technology can reduce costs in the following areas: some production costs for new weapon systems (e.g., through advanced materials); life cycle support systems for new platforms

(e.g., through low-failure design and manufacture); the system development process itself; and training of personnel to use these systems.

20. In addition to its ongoing use of simulation technology in training individual soldiers and small units, the Army needs to explore the technological opportunities for use of simulation systems in R&D and operational evaluation as well as large-scale (i.e., multiple unit) training exercises. In particular, a facility is needed where the complex interactions of the modern mobile battlefield can be simulated with a high degree of realism. Such a facility could be used to evaluate tactics for imminent contingencies and to assess the implications of new technology for tactics, doctrine, and related systems.

21. For more efficient exploitation of advanced technology, the Army can improve its current requirements process by (1) expanding the top-down definition of its role in joint contingency warfare, with an emphasis on how the Army can rely on other services for support and what obligations it may in turn incur; (2) increased early experimental examination of capability options and their costs; and (3) closer, and more frequent, balancing of user needs with technology availability. Among the means to achieve the last two objectives are expanded use of early prototyping and the ATTD program, provided the results can be obtained early enough to contribute to the design and concept formulation processes.

22. The quality of the technologists and acquisition specialists that the Army can recruit, train, and retain will, in the end, determine the Army's ability to participate in the technological revolution foreseen by the STAR Committee and the STAR panels. There is some evidence that technologists in the Army community remain dissatisfied with the work environment despite recent attempts to improve it. However, recent experiences within the DOD Laboratory Demonstration Program and within certain high-quality Army laboratories hold promise for procedural changes that could significantly improve the work environment.

Force Structure and Strategy

23. The Army's immediately deployable forces will need the capability to counter potentially superior numbers, air and missile attack, and heavy armor. They will need weapons with longer reach and more combat power without sacrificing rapid deployment capability. Systems to be deployed with these forces must be transportable by air. The focal value of *deployability* was important in the STAR Committee's selection of high-payoff systems. In the future,

the means of deploying a system will be crucial to its effectiveness and must be addressed early in the design process.

24. An integrated combat capability will be needed to support initially deployed Army forces. In particular, their ability to sustain themselves until heavy forces can be inserted will depend on close coordination with supporting Air Force and Naval forces. Therefore, the STAR-recommended focal value of *joint operability* will be essential to force structure planning.

25. Deployed forces will need to maintain an overwhelming advantage in air superiority and the stand-off capability of ground weapons. The one-sidedness of these advantages can be maintained only if these forces have effective countermeasures to the anticipated increase in use of stealth by adversaries. Counterstealth capabilities will therefore become increasingly valuable in lessening the vulnerability of U.S. support and combat forces deployed for contingency operations.

26. Smart weapons and countermeasures to them will increasingly define the character of ground warfare. Smart weapons can enhance the reach and effectiveness of combat forces. By substituting for tons of dumb steel, they can also lessen the logistics burden of supporting forces deployed a long distance from their bases. However, the effectiveness of smart weapons will depend on more and better C^3I/RISTA as well as force elements well trained in their use.

RECOMMENDATIONS

This section summarizes the various recommendations made throughout this report by the STAR Committee for consideration by the Army. In many cases the rationale to support the recommendation is summarized in one or more of the conclusions presented in the section above.

1. The Army should maintain its current level of support for research and advanced technology (i.e., 6.1, 6.2, and 6.3a funding) despite the expected substantial reductions in overall resources available to its acquisition accounts.

2. The Army should meld into its current Army Technology Base Master Plan the STAR Committee's selection of high-payoff technologies:

- multidomain smart sensor technology;
- terahertz-device electronics;
- secure, wide-bandwidth communications technology;

- battle management software technology;
- solid state lasers and/or coherent diode laser arrays;
- electric-drive technology;
- genetically engineered and developed materials and molecules;
- material formulation techniques for "designer" materials; and
- methods and technology for integrated systems design.

3. The Army should aim to increase the effectiveness of early technology explorations by focusing them on advanced systems concepts. Among these focal interests for technology exploration should be the six systems concepts selected by the STAR Committee:

- robot vehicles (air or ground) for C^3I/RISTA;
- electronic systems architecture;
- brilliant munitions for attacking ground targets;
- lightweight indirect-fire weapons;
- theater air and missile defense; and
- simulation systems for R&D, analysis, and training.

The Army should augment its Technology Base Master Plan with explicit, high-visibility programs for each systems area of focal interest, much as it has already done for the Soldier-as-a-System program. An independent review team should assess progress in each area. Also, a process for adding to or subtracting from the list of priority systems concepts should be established.

4. The focal values of affordability, reliability, deployability, joint operability, reduced vulnerability of U.S. combat and support systems, casualty reduction, and support system cost reduction should be stressed throughout the Army's technology programs. The review of progress in each systems area should also assess performance with respect to these focal values.

5. The Army should implement an expanded test program to examine the potential battlefield impact of both high-payoff technologies and the high-payoff systems into which they might be incorporated. The Army should consider application of force structure assets to this design support and evaluation role, as the Navy does with its VX squadrons.

6. The Army should commit to upgrading the combat capabilities of its first-to-be-deployed light forces and to substantially reducing the weight of systems for its heavy forces, so that a suitable middle tier of medium air-deployable forces can be achieved. Current light forces need, and can be given, more tactical mobility, more survivability, and, especially, more firepower with greater lethality against

hard targets. The lightweight indirect-fire weapon system discussed in this report illustrates the kind of systems concept that is needed and the technology that can help produce it. Weight reduction of heavy-force systems will require, first, a commitment to achieve this goal and, second, applications of new technologies (e.g., advanced materials and propulsion systems) to new designs in which weight ceilings are lowered. The main battle tank could well be the first system to be substantially lightened.

7. The Army should allocate the predominant share of its technology resources to areas not well supported by private sector commercial development. On the other hand, wherever possible, it should rely on commercially derived technologies, components, products, and manufacturing. A policy that outlines these twin approaches and the procedures for their implementation should be rapidly developed and promulgated. Several pilot programs should be initiated to "wring out" these procedures.

8. The Army should increase its use of procedures—such as rapid austere prototyping and subsystem upgrades—that can expedite the movement of technology from the laboratory into the hands of its forces. Such a policy must recognize both the reduced rate of implementation of completely new platforms and the imperative that deployed Army forces at no time be denied technological superiority. Gradual improvement of fielded designs by subsystem upgrades can move new technology into the field faster than simply waiting for a new platform baseline. The early phases of technology programs should incorporate a "design for change" requirement, so the design can accommodate upgraded components and subsystems after it is fielded.

9. The Army should plan to meet future mobilization requirements in light of expected reduced procurement and war reserve material levels. Planning for surge capacity and reconstitution of forces will increase dependence on commercial parts and manufacturing practices.

10. In areas where the Army has vital interests (such as an advanced C^3I network, deployment of forces into areas of heavy armored resistance, and theater defense against air or missile attack), the Army should take the lead in joint planning. A key area for such leadership is coordination of theater air and missile defense systems through the SDIO and other channels.

11. There are many opportunities for improving joint operations with the other services. Most of these require joint consideration and program initiatives during the research, development, and requirements definition phases for new programs. Future Army performance

depends on seizing these opportunities. The Army should participate actively in developing joint program plans and take the lead, if necessary, for at least the following areas:

- providing airlift and sealift for initial forces;
- C^{3}I/RISTA systems;
- air and missile defense, including defensive low-observability, defense against stealthy attack, and IFFN; and
- close air support.

12. The Army should implement an aggressive program to ensure that it will continue to attract, train, and retain people of the highest quality in its advanced technology structure. To that end, the Army should review the results of the DOD Laboratory Demonstration Program and the results of innovative actions already taken in its highest-quality laboratories. Efforts should be made to work cooperatively with civilian and commercial entities to maintain skills and technology transfer. A good example is the placement of Army medical personnel in civilian trauma centers.

13. The Army should modify the implementation of its CBRS. The concept input to the process should allow for greater technology exploration and consideration of potential threats and advanced systems. Rather than assuming that every requirement entering Phase 1 is destined for Phase 4 development, Phase 1 should be more open, with a winnowing process occurring at each move to a subsequent phase. Test and evaluation need to be rethought as tools for learning, and redesigning, from experiment, as in the methodology of rapid austere prototyping. To make these alterations work, a strategic vision for the Army's program must be communicated from the top.

Appendixes

Appendix A

Comparison of Technology Lists

Several lists of defense-related technologies are currently in circulation. Naturally enough, these lists will be compared with one another by the defense community, industry, the academic R&D community, and many others interested in government policy and support for science and technology. To make such comparisons meaningful, two points must be borne in mind:

- There is no standard taxonomy for classifying technologies. Nor is there a recognized and uniform distinction drawn between a technology and an application or systems concept that uses technology.
- The selection criteria for these lists may differ. Indeed, differences should be expected if the purposes of the lists differ.

To compare lists in a meaningful way, it is therefore necessary to go beyond the listed items to the accompanying narrative to determine what each listed item signifies. This appendix will discuss three technology lists, not to draw any conclusions about their merit relative to the STAR list but rather to illustrate the importance of these two points. The following three lists will be considered:

- the list of 13 *key emerging technologies* from the second edition (November 1990) of the Army Technology Base Master Plan;
- the list of 21 Defense Critical Technologies from *The Department of Defense Critical Technologies Plan,* prepared by the Department of Defense for the Committees on Armed Services of the U.S. Congress; and

• the list of 22 National Critical Technologies included in the March 1991 *Report of the National Critical Technologies Panel.*

STAR TECHNOLOGY AND SYSTEMS LISTS

A summary of the selection criteria and the classification approach used in the STAR study is a useful starting point for these comparisons. The STAR Science and Technology Subcommittee divided all Army-related technologies into the eight area-specific technology groups. Each Technology Group used its own classification of technologies it considered to be within its scope. Over a hundred of these "technology species" were identified as having importance to ground warfare in the twenty-first century, which was the key objective specified for the STAR study. In Chapter 3 this full set of technologies is represented by "genus-and-species" short descriptions in the TFA Scope section for each report.

In response to an Army request for a short list of the highest-payoff technologies, representatives of the technology groups were asked to nominate, from each group's scope, one or two technologies that would have the highest technological and operational potential for Army applications. The candidate technologies were then reviewed by the Subcommittee as a whole. The resulting list of nine high-payoff technologies (listed in Table A-1) is relatively exclusive, in that many "species" covered by each group were necessarily omitted.

In addition to the list of nine high-payoff technologies, the STAR Committee has also proposed five high-payoff systems concepts and seven technology-related focal values that pertain to many systems. Depending on the system, a number of technologies—often from different TFA areas—contribute to a particular focal value. Consider, for example, the range of technologies relevant to affordability, to stealth and counterstealth capabilities, or to casualty reduction.

ARMY TECHNOLOGY BASE:
KEY EMERGING TECHNOLOGIES

The first column of Table A-2 lists the 13 key emerging technologies in the Army Technology Base Master Plan (November 1990 edition). The Army described them as "those technologies whose development is considered most essential to ensure the long-term qualitative superiority of Army weapon systems" (U.S. Army, 1990). The technologies share the following characteristics with respect to their value to the Army, mode of selection, and anticipated advances:

TABLE A-1 STAR Technology-Relevant Lists

STAR Technology Groups

1. Computer Science, Artificial Intelligence, and Robotics
2. Electronics and Sensors
3. Optics, Photonics, and Directed Energy
4. Biotechnology and Biochemistry
5. Advanced Materials
6. Propulsion and Power
7. Advanced Manufacturing
8. Environmental and Atmospheric Sciences
9. Basic Sciences (became Long-Term Forecast of Research)

High-Payoff Technologies

1. technology for multidomain smart sensors
2. terahertz device electronics
3. secure, wide-band communications technology
4. battle management software technology
5. solid state lasers and/or coherent diode laser arrays
6. genetically engineered and developed materials and molecules
7. electric drive technology
8. material formulation techniques for "designer" materials
9. methods and technology for integrated systems design

High-Payoff Systems

1. robot vehicles (air or ground) for C^3I/RISTA
2. electronic systems architecture
3. brilliant munitions for attacking ground targets
4. lightweight indirect-fire weapons
5. integrated theater air/missile defense
6. simulation systems for R&D, analysis, and training

Cross-Cutting Focal Values

1. affordability
2. reliability
3. deployability
4. joint operability
5. reduced vulnerability of U.S. combat and support systems (stealth and counterstealth capability)
6. casualty reduction
7. support system cost reduction.

TABLE A-2. Army Technology Base Key Emerging Technologies

Key Emerging Technology	Technology Areas (Selected)	Relevant STAR TFA or Section	STAR High-Payoff Technologies
1. Advanced Materials and Materials Processing	Composite armor; structural composites; high-temperature engine components; soldier body armor	Advanced Materials	material formulation techniques for "designer" materials
2. Microelectronics, Photonics, and Acoustics	VHSIC silicon devices; MMICs; compound semiconductors; photonic devices; fiber-optic sensors; multispectral sensors; focal plane arrays; infrared sensors	Electronics and Sensors; Optics, Photonics, and Directed Energy	multidomain smart sensors; terahertz-device electronics; secure, wideband communications technology
3. Advanced Signal Processing and Computing	Software producibility & life cycle; data base management systems; parallel systems; algorithms; automated systems (voice recognition)	Computer Science, Artificial Intelligence, and Robotics;	methods & techniques for integrated systems design; battle management software technology
4. Artificial Intelligence	High-speed computation; knowledge acquisition & learning; neural nets; multisensor data fusion; adaptive control & robotics; AI for logistics, planning, simulation, maintenance, language training.	Computer Science, Artificial Intelligence, and Robotics; Electronics and Sensors; Optics, Photonics, Directed Energy	terahertz-device electronics; battle management software technology
5. Robotics	sensors; unmanned ground vehicles; data rate reduction; environmental perception; robot manipulators; various robotics applications	Computer Science, Artificial Intelligence, and Robotics; Electronics and Sensors	

6. Advanced Propulsion Technology	small turbine IHPTET engine cores; VTOL aircraft; ground vehicle transmissions; weapon chemical propulsion	Propulsion and Power	electric-drive technology
7. Power Generation, Storage, and Conditioning	energy storage for pulse power; electric drive power conditioning; fuel cells	Propulsion and Power	
8. Directed Energy	laser efficiency; high energy-density capacitors; switches; high-power microwave	Propulsion and Power	solid state lasers pumped by diode lasers
9. Biotechnology	biosensors and enzyme decontamination for CTBW; vaccines; artificial tissues; multivalent assays for disease; bio-remediation; food preservation and packing	Biotechnology and Biochemistry	genetically engineered and developed materials and molecules
10. Space Technology	communications (man-portable, LIGHTSAT; RISTA; terrain & weather; position & navigation; computation for space systems; fire support	Electronics and Sensors; Environmental & Atmospheric Sciences; Optics, Photonics, and Directed Energy	methods & techniques for integrated systems design;
11. Low-Observable Technology	radar: absorbing materials and cross-section reduction; infrared: special coatings, vehicle cooling techniques; visual: coatings to suppress or vary reflectance; noise suppression	Electronics and Sensors Advanced Materials	terahertz-device electronics; material formulation techniques for "designer" materials; methods & techniques for integrated systems design;

TABLE A-2. Army Technology Base Key Emerging Technologies (continued)

Key Emerging Technology	Technology Areas (Selected)	Relevant STAR TFA or Section	STAR High-Payoff Technologies
12. Protection/Lethality	armor; soldier eye protection; kinetic energy projectiles; explosives; mine detection; CTBW detection, protection, decontamination	Advanced Materials; Propulsion and Power; Biotechnology	material formulation techniques for "designer" materials; genetically engineered and developed materials and molecules
13. Neuroscience Technology	sleep studies; physiological response to adverse environments; sensory-motor integration applications to advanced weapons systems, robotics	See Systems Panel Reports on Health and Medicine; Personnel; Special Technologies	

• All hold great promise for solving important deficiencies or significantly increasing U.S. capabilities on the modern battlefield.

• All were reviewed by Army technical managers, scientists, and engineers in terms of future needs, then presented to users and developers and finally approved by the Army leadership.

• Each technology is promising but still immature. Knowledge gaps must be filled for each before technical decisions can be made about its use in Army applications.

Except for the mode of selection (item 2), these characteristics seem reasonably similar to those used by the STAR study to select the high-payoff technologies. However, these key emerging technologies represent a far broader classification than the STAR high-payoff technologies. Indeed, the first nine correspond quite closely with entire STAR TFAs or with major sections from several TFAs. (Compare columns 1 and 3 in Table A-2.)

The level of technology classification in the Army Technology Base Master Plan that corresponds better to the STAR technologies consists of the technology areas. These are listed in milestone tables for each key emerging technology (See U.S Army, 1990, Volume I, Chapter III). A representative sample of these technology areas is shown in column 2 of Table A-2.

There are more than 70 technology areas for the 13 key emerging technologies. This is the same order of magnitude as the hundred-odd technologies identified by the STAR Science and Technology Subcommittee. A comparison of the technology areas with the STAR technology species summarized in Chapter 3 shows them to be at a roughly equivalent level of detail. (Close comparison is sometimes difficult because some technology areas are application-oriented, which makes them closer to what STAR calls advanced systems concepts. The relation to similar STAR technology species is often clearer from the milestone descriptions that accompany each technology area.)

Of the four remaining key emerging technologies, three represent classifications by application: space technology, low-observable technology, and protection/lethality. Space technology and protection/lethality were addressed by STAR through its systems concepts. (See Space-Based Systems and the sections on lethal systems in Chapter 2; see also the recommendations for these system areas in Chapter 5.) Low-observable technology is represented by the STAR focal value of stealth and counterstealth capabilities; it is cited as an important feature of many STAR systems concepts and as a value to which many of the STAR-forecast technology advances could contribute.

The last of the key emerging technologies, neurosciences, comprises technology areas that correspond to several systems concepts described by STAR for Integrated Soldier Support (see Chapter 2), as well as technology species assessed by the Biotechnology and Biochemistry Group (biocoupling and bionics) and by the Computer Science, Artificial Intelligence, and Robotics Group (sensor-motor integration in advanced robot systems).

In summary, a meaningful comparison of the STAR study with the Army Technology Base technologies should compare nine of the key emerging technologies with the full range of six of the STAR TFAs, not just the nine high-payoff technologies. Three of the key emerging technologies are best compared with systems concepts developed by the STAR systems panels. The key emerging technologies are in fact an *inclusive classification* rather than a highly specific and *exclusive* selection like the STAR high-payoff technologies. The fourth column of Table A-2 shows where the STAR high-payoff technologies and systems fall with respect to this "key emerging technologies" classification.

The point is not that one or the other list is better but rather that they are not directly comparable. In fact, if this set of Army Technology Base key emerging technologies seems particularly well suited to the Army's needs, perhaps it should be used as the starting point for the general technology classification to be used in future studies analogous to that of STAR. On the other hand, the key emerging technologies are so encompassing that, as a list, they do not provide specific input for focusing Army R&D efforts. An action-oriented list would need to be at the level of technology areas rather than key emerging technologies.

DEFENSE CRITICAL TECHNOLOGIES

Table A-3 lists the 21 Defense Critical Technologies and, in the second column, their principal component fields as specified in the original report (DOD, 1991). For a number of these technologies, the extent of their component fields shows that they are close in scope to an entire STAR TFA or to major sections of several TFAs. Compare, for example, the component fields of Semiconductor Materials and Microelectronic Circuits with the TFA scope of Electronics and Sensors or the fields for Composite Materials with the TFA scope for Advanced Materials.

Some of the Defense Critical Technologies do correspond to just a few closely related STAR technology species. Two examples are Pulsed Power (compare with the technologies for *pulsed and short-*

TABLE A-3. Defense Critical Technologies Compared with STAR

Defense Critical Technologies	Principal Component Fields (Selected)	Related STAR TFAs (and Systems Panels)	STAR High-Payoff Technologies and Systems
1. Semiconductor Materials & Microelectronic Circuits	VLSICs; CAD for complex circuits; high-resolution lithography; A/D converters; MMICs; T/R modules; signal control components; radiation-hard isolation	Electronics and Sensors	
2. Software Engineering	system engineering process & environment; fault-tolerant software; software for parallel & heterogeneous stems;	Computer Science, Artificial Intelligence, and Robotics	methods & techniques for integrated systems design; electronic systems architecture
3. High Performance Computing	Advanced software technology & algorithms; networking; computer sci. research personnel; high-density packaging;	Computer Science, Artificial Intelligence, and Robotics	
4. Machine Intelligence & Robotics	image understanding; autonomous planning; navigation; speech & text processing; machine learning; knowledge representation and acquisition; adaptive manipulation & control	Computer Science, Artificial Intelligence, and Robotics	battle management software technology; robot vehicles (air or ground) for C³I/RISTA
5. Simulation & Modeling	high-speed graphics; solutions for nonlinear equations; simulation verification and validation	Computer Science, Artificial Intelligence, and Robotics	simulation systems for R&D, analysis, and training

TABLE A-3. Defense Critical Technologies Compared with STAR (continued)

Defense Critical Technologies	Principal Component Fields (Selected)	Related STAR TFAs (and Systems Panels)	STAR High-Payoff Technologies and Systems
6. Photonics	laser devices; fiber optics; optical signal processing; integrated optics	Optics, Photonics, and Directed Energy	solid state lasers pumped by laser diodes; secure wideband communications; multidomain smart sensors
7. Sensitive Radar	Advanced monostatic radar; multistatic radar; radars for target recognition; laser radar; electronic counter countermeasures	Electronics and Sensors; Optics, Photonics, and Directed Energy; see also Electronics Systems	terahertz electronic devices; integrated theater air/missile defense
8. Passive Sensors	thermal imagers; IR focal plane arrays; IR search and track; diffractive optics; sensor integration; passive antennas; passive RF and acoustic surveillance; fiber-optic sensors for environmental and systems monitoring; superconducting sensors	Optics, Photonics, and Directed Energy; Electronics and Sensors; Advanced Materials; see also Electronics Systems;	multidomain smart sensors; solid state lasers pumped by diode lasers; secure wideband communications; material formulation techniques for "designer" materials; methods & techniques for integrated systems design;
9. Signal and Image Processing	algorithm development; hybrid optical-digital techniques; phased array control; neural networks	Computer Science, Artificial Intelligence, and Robotics; Optics, Photonics, and Directed Energy; Electronics Systems	electronic systems architecture; methods & techniques for integrated systems design

10. Signature Control	radar cross-section reduction; IR, visual, UV, magnetic signature reduction/management; acoustic quieting; LPI radar, communications, and navigation;	Advanced Materials; Electronics and Sensors; Optics, Photonics, and Directed Energy; Electronics Systems; Airborne Systems	secure wideband communications; material formulation techniques for "designer" materials; (see also stealth-counterstealth focal value)
11. Weapon System Environment	Environment characterization and prediction; target environment analysis and simulators	Environmental and Atmospheric Sciences; Computer Science, Artificial Intelligence, and Robotics	simulation systems for R&D, analysis, and training; methods & techniques for integrated systems design
12. Data Fusion	theory; algorithms and models; data base and knowledge base for fusion; reasoning systems	Computer Science, Artificial Intelligence, and Robotics; Electronics and Sensors; Optics, Photonics and Directed Energy	multidomain smart sensors; terahertz device electronics; secure wideband communications; battle management software technology
13. Computational Fluid Dynamics (CFD)	Unsteady aerodynamic regimes; turbulence modeling; internal flows	Propulsion and Power; see also Airborne Systems Panel Report	
14. Air-Breathing Propulsion	high pressure ratio, lightweight compression systems; high-temperature, improved-life combustion systems; high efficiency turbines; reduced-signature multifunctional nozzles; adaptive, survivable, high-speed integrated control systems; scramjet technology;	Propulsion and Power; see also Airborne Systems Panel Report	methods and technology for integrated systems design

TABLE A-3. Defense Critical Technologies Compared with STAR (continued)

Defense Critical Technologies	Principal Component Fields (Selected)	Related STAR TFAs (and Systems Panels)	STAR High-Payoff Technologies and Systems
15. Pulsed Power	Energy storage; power switching and conditioning; high-power microwave	Propulsion and Power; Optics, Photonics and Directed Energy	solid state lasers and/or coherent diode-laser arrays
16. Hypervelocity Projectiles and Propulsion	projectile design; propulsion; projectile-target interaction	Propulsion and Power; see also Lethality Systems	integrated theater air/missile defense
17. High Energy-Density Materials	explosives; propellants; warheads; nuclear isomers	Propulsion and Power; Advanced Materials	lightweight indirect-fire weapons
18. Composite Materials	organic matrix composites; metal matrix composites; ceramic matrix composites, hybrid composites	Advanced Materials	material formulation technologies for "designer" materials
19. Superconductivity	low-temperature and high-temperature superconductors for magnets, sensors, electronics	Electronics and Sensors	
20. Biotechnology		Biotechnology and Biochemistry	genetically engineered and developed materials and molecules
21. Flexible Manufacturing	computer-aided design, engineering, manufacturing, production; data bases; communications and networking; intelligent software interfaces	Computer Science, Artificial Intelligence, and Robotics; Advanced Manufacturing	methods & techniques for integrated systems design

duration power and *power conditioning* under Battle Zone Electric Power in the Propulsion and Power section of Chapter 3) and High Energy Density Materials (see Energetic Materials in the Advanced Materials section). Still other technologies on this list correspond more closely to STAR advanced systems concepts than to STAR technology classifications. Examples are Sensitive Radar, Passive Sensors, or Weapon System Environment.

Again, the point of this comparison is not to determine which classification is better but to show that the Defense Critical Technologies in fact cover a broad area roughly coextensive with the entire set of STAR TFAs, plus some of the technology applications (systems concepts) covered by the STAR systems panels.

NATIONAL CRITICAL TECHNOLOGIES

The National Critical Technologies (NCT) Panel was charged with identifying up to 30 national critical technologies in its biennial report to the President and Congress. A national critical technology is defined as an area of technological development that is essential for the long-term national security and economic prosperity of the United States (National Critical Technologies Panel, 1991). Unlike the STAR study or the other two list-producing activities discussed above, the NCT Panel is responsible for technology that is nonmilitary but nonetheless essential to economic prosperity. Thus, its list can be expected to have a wider compass than the defense-related classifications.

Table A-4 lists the 22 NCTs in column 2, with the NCT class headings shown in the first column. Although this table does not summarize the description of each technology from the NCT report, the technology titles show that this list, too, represents a broader level of technology aggregation than most of the STAR high-payoff technologies. The correlation to STAR TFAs or major sections of TFAs is quite high (column 3 of Table A-4), especially in light of the broader ambit of the NCT Panel's mission.

CONCLUSIONS

Each of the three technology lists is more meaningfully compared with the combined scopes of the STAR TFAs than with the STAR high-payoff technologies. The latter were intended to be a fairly exclusive handful of specific rapidly advancing technologies whose pursuit seems most likely to produce major technology payoffs for the Army.

TABLE A-4. National Critical Technologies and STAR Technologies

NCT Category	National Critical Technologies	Related STAR TFAs	STAR High-Payoff Technologies and Systems
Materials	Materials synthesis and processing	Advanced Materials	material formulation techniques for "designer" materials
	Electronic and photonic materials	Electronics and Sensors; Optics, Photonics and Directed Energy	terahertz-device electronics
	Ceramics	Advanced Materials	
	Composites	Advanced Materials	
	High-performance metals and alloys	Advanced Materials	
Manufacturing	Flexible computer-integrated manufacturing	Advanced Manufacturing; Computer Science, Artificial Intelligence, and Robotics	methods and technology for integrated system design
	Intelligent processing equipment	Advanced Manufacturing; Computer Science, Artificial Intelligence, and Robotics	
	Microfabrication, nanofabrication	Advanced Manufacturing; Electronics and Sensors	
	Systems management technologies	Advanced Manufacturing; Computer Science, Artificial Intelligence, and Robotics	

Information and Communications	Software	Computer Science, Artificial Intelligence, and Robotics	methods and technology for integrated system design; battle management software technology
	Microelectronics and optoelectronics	Electronics and Sensors; Optics, Photonics, and Directed Energy	terahertz-device electronics
	High-performance computing and networking	Computer Science, Artificial Intelligence, and Robotics	electronic systems architecture
	High-definition imaging and displays	Electronics and Sensors; Optics, Photonics, and Directed Energy	
	Sensors and signal processing	Electronics and Sensors; Optics, Photonics, and Directed Energy	multidomain smart sensors; terahertz-device electronics; secure wideband communications
	Data storage and peripherals	(not specifically addressed in STAR TFAs)	electronic systems architecture
	Computer simulation and modeling	Computer Science, Artificial Intelligence, and Robotics	simulation systems for R&D, analysis, and training
Biotechnology and Life Sciences	Applied molecular biology	Biotechnology and Biochemistry	
	Medical technology	(Health and Medical Systems Panel)	

TABLE A-4. National Critical Technologies and STAR Technologies (continued)

NCT Category	National Critical Technologies	Related STAR TFAs	STAR High-Payoff Technologies and Systems
Aeronautics and Surface Transportation	Aeronautics	Propulsion and Power; Airborne Systems Panel	robot vehicles (air or ground) for C³/RISTA
	Surface transportation technologies	Propulsion and Power	electric drive technology
Energy and Environment	Energy technologies	Propulsion and Power; Advanced Materials	material formulation techniques for "designer" materials
	Pollution minimization, remediation, and waste management	Biotechnology and Biochemistry; (see also Health and Medical Systems Panel)	

On the other hand, when the items on these lists are understood in terms of the scopes specified for them by the lists' authors, there is surprisingly wide agreement among them and with the full set of important technologies represented in the STAR TFAs and reports of the Systems Panels.

REFERENCES

DOD. 1991. The Department of Defense Critical Technologies Plan for the Committees on Armed Services, United State Congress. Department of Defense Report AD-A234 900. Washington, D.C. May 1.

National Critical Technologies Panel. 1991. Report of the National Critical Technologies Panel. Superintendent of Documents, U.S. Government Printing Office. Washington, D.C. March.

U.S. Army. 1990. Army Technology Base: Volume 1. Headquarters, Department of the Army, Deputy Assistant Secretary for Research and Technology (SARD-ZT). Washington, D.C. November.

Appendix B

Contributors to the STAR Study

COMMITTEE ON STRATEGIC TECHNOLOGIES FOR THE ARMY (STAR)

Executive Committee

Willis M. Hawkins, (General Study Chairman), Lockheed
  Corporation (Retired)
Robert R. Everett, The MITRE Corporation (Retired)
Martin A. Goland, Southwest Research Institute
Ray L. Leadabrand, Leadabrand and Associates
Michael D. Rich, The RAND Corporation

Army Liaison Personnel to Executive Committee

Richard Chait, U.S. Army Materiel Command

Integration Subcommittee

Ray L. Leadabrand (*Chairman*), Leadabrand and Associates
David D. Elliott, Systems Control Technology, Inc.
David C. Hazen, Princeton University (Professor Emeritus)
Walter B. LaBerge, Lockheed Corporation (Retired)
GEN John W. Pauly, Systems Control Technology, Inc.
Charles J. Shoens, Science Applications International Corporation
  (Retired)

Army Liaison Personnel to
Integration Subcommittee

LTC Albert Sciarretta, U.S. Army Materiel Command
James Fox, CACDA

Technology Management and Development Planning Subcommittee

Michael D. Rich (Chairman), The RAND Corporation
Walter B. LaBerge, Lockheed Corporation (Retired)
VADM William J. Moran, Consultant
GEN John W. Pauly, Systems Control Technology, Inc.
GEN John W. Vessey, Jr., Consultant

Science and Technology Subcommittee

Robert R. Everett (Chairman), The MITRE Corporation (Retired)
John B. Wachtman, Jr. (*Vice Chairman*), Rutgers University
W. James Sarjeant (*Special Technical Advisor*), State University of New
   York at Buffalo

Army Liaison Personnel to Science and
Technology Subcommittee

Richard E. Smith, U.S. Army Materiel Command

Senior Technical Council

Donald S. Fredrickson, Consultant
Edward A. Frieman, Scripps Institution of Oceanography
George E. Solomon, TRW (Retired)
John B. Wachtman, Jr., Rutgers University

Technology Groups of the Science and Technology Subcommittee

Advanced Manufacturing Technology Group

James G. Ling (*Lead Expert*), Consultant
Roger N. Nagel (*Deputy Lead Expert*), Lehigh University
Theodore W. Schlie, Lehigh University

Army Liaison Personnel

Steve V. Balint, U.S. Army Materiel Command

Advanced Materials Technology Group

John D. Venables (*Lead Expert*), Martin-Marietta Corporation (Retired)
Yet-Ming Chiang, Massachusetts Institute of Technology
Lawrence T. Drzal, Michigan State University
John J. Lewandowski, Case Western Reserve University

Army Liaison Personnel

Robert N. Katz, Worcester Polytechnic Institute
Joseph A. Lannon, U.S. Army Armament, Munitions and Chemical
    Command

Basic Sciences Technology Group

Gordon J. F. MacDonald (*Lead Expert*), University of California,
    San Diego
Jonathan I. Katz, Washington University

Army Liaison Personnel

Roy Roth, U.S. Army Research Office

Biotechnology and Biochemistry Technology Group

Christopher C. Green (*Lead Expert*), General Motors Research Labs
Joseph F. Soukup (*Vice Lead Expert*), Science Applications
    International Corporation
Bertram W. Maidment, Jr., Midwest Research Institute
Gregory A. Petsko, Brandeis University

Army Liaison Personnel

Daphne Kamely, Office of the Assistant Secretary of the Army,
    Research, Development and Acquisitions

Computer Science, Artificial Intelligence, and Robotics Technology Group

Ruth M. Davis (*Lead Expert*), Pymatuning Group
Allen C. Ward (*Vice Lead Expert*), University of Michigan
David H. Brandin, A. T. Kearney Technology, Inc.
Brian P. McCune, Booz, Allen & Hamilton, Inc.

Army Liaison Personnel

Harold Breaux, U.S. Army Ballistic Research Laboratory
David C. Hodge, U.S. Army Human Engineering Laboratory
Ronald Hofer, U.S. Army Project Manager for Training Devices
Som D. Karamchetty, U.S. Army Laboratory Command

Electronics and Sensors Technology Group

Walter E. Morrow, Jr. (*Lead Expert*), Massachusetts Institute of
    Technology
Thomas C. McGill, Jr., California Institute of Technology
Edwin B. Stear, The Boeing Company

Army Liaison Personnel

Gerald J. Iafrate, U.S. Army Research Office
Clarence Thornton, Electronics Technology and Devices Laboratory

Environmental and Atmospheric
Sciences Technology Group

George F. Carrier (*Lead Expert*), Harvard University
Wilbert Lick, University of California at Santa Barbara
Edward M. Mikhail, Purdue University

Army Liaison Personnel

James E. Morris, U.S. Army Atmospheric Science Laboratory

Long-Term Forecast of Research
Workshop Participants

Dr. John B. Wachtman, Jr., (*Workshop Chairman*), Rutgers University
Robert R. Everett, The MITRE Corporation (Retired)
J. Christian Gillin, San Diego Veterans Administration Medical
    Center
John B. Harkins, Texas Instruments
Walter B. LaBerge, Lockheed Corporation (Retired)
Wilbert Lick, University of California at Santa Barbara
Edward M. Mikhail, Purdue University
Gregory A. Petsko, Brandeis University
Richard S. Shevell, Stanford University
Joseph F. Soukup, Science Applications International Corporation

John D. Venables, Martin-Marietta Corporation (Retired)
Allen C. Ward, University of Michigan

Army Liaison Personnel

Gerald Iafrate, Director, Army Research Office (ARO), Research
    Triangle Park, North Carolina
Robert J. Campbell, Chemical and Biological Sciences, ARO
Jagdish Chandra, Director, Mathematical Sciences Division, ARO
Joseph Del Vecchio, Director, Space Programs Laboratory, U.S. Army
    Engineering Topographic Laboratories, Fort Belvoir, Virginia
Walter Flood, Director, Geosciences, ARO
Robert Ghirardelli, Director, Chemical and Biological Sciences, ARO
Robert Guenther, Director, Physics Division, ARO
James Mink, Director, Electronics Division, ARO
John Praeter, Associate Director, Material Sciences, ARO
Roy Roth, Technical Support Office, ARO
Michael Stroscio, Senior Scientist, ARO

Optics, Photonics, and Directed Energy
Technology Group

Alfred B. Gschwendtner (*Lead Expert*), Massachusetts Institute of
    Technology
Louis C. Marquet, Nichols Research Corporation
Shen Y. Shey, Massachusetts Institute of Technology
Richard C. Williamson, Massachusetts Institute of Technology

Army Liaison Personnel

Rudolph Buser, U.S. Army Communications—Electronics Command
Donald R. Ponikvar, Office of the Secretary of Defense
John Pollard, CECOM Center for Night Vision and Electro-optics

Propulsion and Power Technology Group

Gerard W. Elverum (*Lead Expert*), TRW, Inc. (Retired)
W. James Sarjeant (*Vice Lead Expert*), State University of New York
    at Buffalo

Army Liaison Personnel

William D. Stephens, U.S. Army Missile Command

Systems Panels

Airborne Systems Panel

Richard S. Shevell (*Chairman*), Stanford University
Basil S. Papadales (*Vice Chairman*), W. J. Schafer Associates
John F. Cashen, Northrup Corporation
Wesley L. Harris, The University of Tennessee Space Institute
Ira F. Kuhn, Jr., Directed Technologies, Inc.
Anthony F. LoPresti, Loral Aeroneutronics
Robert M. McKillip, Jr., Princeton University
Robert H. Widmer, General Dynamics (Retired)

Ex-Officio Member

Lawrence T. Drzal, Michigan State University

Army Liaison Personnel

Bruce W. Fowler, U.S. Army Missile Command
William McCorkle, U.S. Army Missile Command
David J. Weller, U.S. Army Aviation Systems Command

Electronics Systems Panel

J. Fred Bucy (*Chairman*), Texas Instruments Incorporated (Retired)
John K. Harkins (*Vice Chairman*), Texas Instruments Incorporated
William Kelly Cunningham, LTV Missiles
Mike W. Fossier, Raytheon Company
Ronald J. Marini, Martin Marietta Electronic Systems
Richard T. Roca, AT&T Bell Laboratories
Robert A. Singer, Hughes Aircraft Company
Ronald D. Sugar, TRW Space Communications Division
Richard P. Wishner, Advanced Decision Systems

Ex-Officio Members

Yet-Ming Chiang, Massachusetts Institute of Technology
Alfred B. Gschwendtner, Massachusetts Institute of Technology
Edwin B. Stear, The Boeing Company
Allen C. Ward, University of Michigan

Army Liaison Personnel

Anthony V. Campi, U.S. Army Electronics—Communications Command

Woodrow C. Holmes, U.S. Army Electronics—Communications Command

Eugene Famolari, Jr., U.S. Army Electronics—Communications Command

Health and Medical Systems Panel

Richard M. Krause (*Chairman*), National Institutes of Health

William R. Drucker (*Vice Chairman*), Uniformed Services University of the Health Sciences

Joseph M. Davie, G.D. Searle & Company

J. Christian Gillin, San Diego Veterans Administration Medical Center

Robert W. Mann, Massachusetts Institute of Technology

Jay P. Sanford, Antimicrobial Therapy, Inc.

Kenneth W. Sell, Emory University School of Medicine

Army Liaison Personnel

COL Steven Hursh, Walter Reed Army Institute of Research

Robert H. Mosebar, U.S. Army Health Services Command

Lethal Systems Panel

Joseph Sternberg (*Chairman*), Naval Postgraduate School

Harry D. Fair, The University of Texas at Austin

Michael E. Melich, Naval Postgraduate School

Hyla S. Napadensky, Napadensky Engineers, Inc.

Walter R. Sooy, Lawrence Livermore National Laboratory

Michael A. Wartell, Sandia National Laboratories

Army Liaison Personnel

John Kramar, U.S. Army Materiel Systems Analysis Activity

Mobility Systems Panel

William G. Agnew (*Chairman*), General Motors Corporation (Retired)
Gary L. Borman, University of Wisconsin
Gordon R. England, General Dynamics
William J. Harris, Texas A&M University
Robert F. Stengel, Princeton University
Robert D. Wismer, John Deere Corporation
Charles D. Wood, Southwest Research Institute

Army Liaison Personnel

Wayne K. Wheelock, U.S. Army Tank-Automotive Command

Personnel Systems Panel

David C. Nagel (*Chairman*), Apple Computer, Inc.
Earl B. Hunt (*Vice Chairman*), University of Washington
James G. Greeno, Stanford University
Alan M. Lesgold, University of Pittsburgh
Martin A. Tolcott, Office of Naval Research (Retired)
Joseph Zeidner, The George Washington University

Army Liaison Personnel

Myron A. Fischl, Office of the Deputy Chief of Staff for Personnel, U.S. Army

Special Technologies Systems Panel

Gilbert F. Decker (*Chairman*), ACUREX Corporation
Robert A. Beaudet, University of Southern California
Lawrence J. Delaney, Montgomery and Associates
Paul G. Kaminski, Technology Strategies and Alliances
Charles A. Rosen, Consultant
Harold F. O'Neil, Jr., University of Southern California
Tito T. Serafini, TRW, Inc.
Clarence H. Stewart, Processing Research, Inc.
Peter J. Weinberger, AT&T Bell Laboratories

Army Liaison Personnel

David Heberlein, U.S. Army Belvoir Research, Development and
Engineering Center
Robert W. Lewis, U.S. Army Natick Research, Development and
Engineering Center

Support Systems Panel

David C. Hardison (*Chairman*), Consultant
Morton B. Berman, The RAND Corporation
Alan B. Perkins, The MITRE Corporation
Richard P. Wishner, Advanced Decision Systems

Ex-Officio Members

John J. Lewandowski, Case Western Reserve University
Edward K. Mikhail, Purdue University
W. James Sarjeant, State University of New York at Buffalo
Allen C. Ward, University of Michigan

Army Liaison Personnel

Clayton "Richard" Lee, U.S. Army Logistics Center

Study Committee on STAR (SCOS)

Walter B. LaBerge (*Chairman*), Lockheed Corporation (Retired)
Lawrence J. Delaney, Montgomery and Associates
David D. Elliott, Systems Control Technology, Inc.
Robert R. Everett, The MITRE Corporation (Retired)
Donald S. Fredrickson, Consultant
Edward A. Frieman, Scripps Institution of Oceanography
Martin A. Goland, Southwest Research Institute
Christopher C. Green, General Motors Research Labs
David C. Hardison, Consultant
Willis M. Hawkins, Lockheed Corporation (Retired)
Ray L. Leadabrand, Leadabrand and Associates
GEN John W. Pauly, Systems Control Technology, Inc.
Michael D. Rich, The RAND Corporation
George E. Solomon, TRW, Inc. (Retired)
GEN John W. Vessey, Jr., Consultant
John B. Wachtman, Jr., Rutgers University

Additional Army Liaison Support

LTC(P) Daniel M. Ferezan, U.S. Army Training & Doctrine
    Command (TRADOC)
John Hansen, U.S. Army Engineering Topographic Laboratories
MG Jerry C. Harrison, U.S. Army Laboratory Command
Cleves H. Howell III, U.S. Army Materiel Command
Peter G. Pappas, U.S. Army Strategic Defense Command
COL Douglas J. Richardson, U.S. Special Operations Command
COL Thomas E. Stalzer, U.S. Army Operational Test and
    Evaluation Agency

STAR Study Staff

Archie L. Wood, interim Study Director
Catharine E. Little, Study Director, April 1990 to April 1991
Kay S. Kimura, Study Director, October 1988 to April 1990
Donald A. Siebenaler, Senior Staff Officer, February 1991 to
    October 1991
Ann M. Stark, Project Officer and Administrative Coordinator
Margo L. Francesco, Administrative Secretary
Kelly Norsingle, Senior Secretary

STAR Editorial Staff

Robert J. Katt, Senior Writer/Editor, Commission on Engineering
    and Technical Systems
Lynn D. Kasper, Technical Editor, Commission on Engineering
    and Technical Systems
Glenn P. Hecton, Technical Writer, Washington, D.C.
Caroletta Lowe, Technical Editor, Ellicott City, Maryland
Katharine Carlson, Technical Editor, Reston, Virginia

Graphic Art

Kay Amundson, Renton, Washington
Leigh Coriale, National Academy Press, Washington, D.C.

BOARD ON ARMY SCIENCE AND TECHNOLOGY*

Martin A. Goland (Chairman), Southwest Research Institute
J. Fred Bucy, Consultant
Richard C. Flagan, California Institute of Technology
M. Frederick Hawthorne, University of California
David C. Hazen, Consultant
Earl B. Hunt, University of Washington
Robert G. Loewy, Rensselaer Polytechnic Institute
William D. Manly, Consultant
Hyla S. Napadensky, Consultant
Peter G. Olenchuk, Major General, U.S. Army (Retired)
Richard M. Osgood, Jr., Columbia University
Joseph Sternberg, Naval Postgraduate School

BOARD STAFF

Bruce Braun, Director, November 1991 to present
Archie L. Wood, interim Director
Catharine E. Little, Acting Director, 1990 to 1991
Donald A. Siebenaler, Senior Staff Officer, February 1991
Helen D. Johnson, Administrative Associate
Ann M. Stark, Administrative Assistant
Margo L. Francesco, Administrative Secretary
Kelly Norsingle, Senior Secretary

*The terms of all members of the Board at the time the STAR study was initiated expired in June 1990.

Glossary

AAE	Army Acquisition Executive, an organizational unit of the Army resulting from the 1989 Army Management Review.
ADC	Analog-to-digital converter.
ASA(RDA)	Assistant Secretary of the Army (Research, Development and Acquisition).
ASIC	Application-specific integrated circuit.
ATC/ATR	Automatic target cueing and recognition.
ATTD	Advanced Technology Transition Demonstration, a development approach used by the Army to test the application of new technology.
BAST	Board on Army Science and Technology (of the National Research Council).
C³I/RISTA	Systems or applications typically included under command, control, communication, and intelligence (C³I) or under reconnaissance, (intelligence), surveillance, and target acquisition (RISTA or RSTA)
CBRS	The Army's Concept-Based Requirements System.
CEP	Circular error probability.
CIM	Computer-integrated manufacturing.
COFT	Conduct-of-fire trainer.
CRAF	Civil Reserve Air Fleet.
CTBW	Chemical, toxin, or biological warfare.
DARPA	Defense Advanced Research Projects Agency.
DBMS	Data base management system.

DNA	Deoxyribonucleic acid.
DOD	U.S. Department of Defense.
DRAM	Direct random access memory.
DSP	Digital signal processing (microprocessor).
ECM/ESM	Electronic countermeasures and support measures.
EHF	Extremely high frequency (radio waves).
EM	Electromagnetic.
ETC	Electrothermal chemical (a gun propulsion technology).
FEL	Free-electron laser.
HALE	High-altitude, long-endurance (aircraft).
HDE	High-power directed-energy (devices or weapons).
HF/DF	Hydrogen fluoride/deuterium fluoride (a chemical laser).
HMX	Current-generation Army explosive (cyclotetramethylene tetranitramine).
HPM	High-power microwave.
HUMINT	Human intelligence.
ICAI	Individual computer-aided instruction.
IFF	Identification of friend or foe.
IFFN	Identification of friend, foe, or neutral.
IHPTET	Integrated High-Performance Turbine Engine Technology (an engine research program).
IPS	Integrated Propulsion System (program).
IRST	Infrared search and track (system).
JSTARS	Joint Systems Target Acquisition Radar System.
LIDAR	Light detection and ranging.
LO	Low observable (also referred to as "stealth"); can refer to technology, designs, etc., that give the capability of being difficult for an enemy to detect or systems that possess this capability.
LPI	low probability of intercept.
MILSATCOM	Military Satellite Communications (a proposed satellite communications architecture).
MLRS	Multiple-launch rocket system.
MMIC	Monolithic microwave integrated circuit.
MOPS	Million (computing) operations per second.
OSD	Office of the Secretary of Defense.
PM/RS	Powder metallurgy for rapidly solidified (alloys).
POMCUS	Prepositioning of materiel configured to unit sets.
PSYOPS	Psychological warfare operations.
R&D	Research and development.
RISTA	Reconnaissance, intelligence, surveillance, and target acquisition; the Army also uses RSTA (see C^3I/RISTA).

RDX	Current-generation Army propellant (cyclotrimethylene trinitramine).
SDI	Strategic Defense Initiative.
SDIO	Strategic Defense Initiative Organization.
SIGINT	Signal intelligence.
SIMNET	Simulation Network.
SQUID	Superconducting quantum interference device.
SRAM	Static random access memory.
STAR	Strategic Technologies for the Army.
TBM	Theater ballistic missile.
TFA	Technology Forecast Assessment.
TMDP	Technology Management and Development Planning (subcommittee of the STAR study).
TRADOC	U.S. Army Training and Doctrine Command.
UAV	Unmanned air vehicle.
UGV	Unmanned ground vehicle.
VLSIC	Very large-scale integrated circuit.
VTOL	Vertical takeoff and landing
WST	Wafer-scale technology.

Index